REVOLVE

REVOLVE
Man's Scientific Rise to Godhood

Aaron Franz

Franz Productions

Copyright © Aaron Franz, 2012

First Published in the United States by Franz Productions, 2010

ISBN 978-0-9853769-0-1

All rights reserved. No part of this publication may be reproduced, stored in a retrieval system, or transmitted in any form or by any means without the prior written permission of the publisher, nor be otherwise circulated in any form of binding or cover other than that in which it is published and without a similar condition being imposed on the subsequent purchaser.

Published by Franz Productions
3410 La Sierra Ave F48
Riverside, CA 92503
franz@theageoftransitions.com

theageoftransitions.com

This book is dedicated to my family, who passed on to me the true riches of the world.
Let us preserve this treasure for future generations.

CONTENTS

Foreword by James Corbett.................................... 9
Introduction... 13

Part One: What is Going On?

1. **The Transhuman Project**................................ 19
   The Singularity... 24
   Alchemy and Transhumanism........................ 31
2. **It's Official**.. 41
3. **Public Relations**... 55
4. **Paradise Engineering**................................... 71
   Virtual Reality... 76
5. **Total Destruction of Privacy**......................... 87
6. **Eugenics**.. 103
   Bringing About Collective Evolution............ 114

Part Two: Digging Deeper

7. **Logic and Reason**.. 127
   The European Enlightenment....................... 136
   Bringing It Full Circle................................... 148
8. **Postgenderism**... 155
   The Artificial Womb..................................... 174
   The Future of Sex... 181
9. **A Worldwide Death and Rebirth Ritual**....... 187
   The Power of Ideas....................................... 203

10. Let's Get Real..	211
Internet Traps...	219
The Dark Light of the New Age............................	227
11. Devil's Advocate..	231
Twisted Virtue..	239
12. Alchemy Throughout the Ages	
The Rock of Ages...	251
The Builder..	258
Solve Et Coagula...	264
Conclusion..	273
Acronym List..	281
Illustration and Graphic Credits................................	283
Bibliography...	284
Index..	290

-Foreword-

By James Corbett, Editor of The Corbett Report - corbettreport.com

What does it mean to be human?

As a young boy that question struck me as the stuff of *Star Trek*, the type of inquiry that would provoke an arched eyebrow from Commander Spock or a head tilt from Lt. Commander Data. Not, in other words, a topic for serious debate.

Growing up in a household dominated by males (myself, two brothers and my father), my poor, beleaguered mother had to put up with all manner of action movies, crude comedies, and of course *Star Trek*, always *Star Trek*.

Like many a young nerd, I was hooked at an early age on this strangely utopian fantasy of a post-racism, post-scarcity, (mostly) peaceful future. And like many a young nerd it instilled in me a fascination with space, a love affair with science, and a conviction that all the world's problems will eventually be solved by technology.

I ended up pursuing that pre-programmed fascination all the way through to university where I began my academic indoctrination as a physics major. Three months later, getting decent grades but hating every minute of it, I changed my major to English literature and thereby decoupled my life forever from my imagined future of lab coats and Bunsen burners.

But the question was to make its reappearance soon enough: What does it mean to be human?

This time it came in the form of an English professor, actually a grad student who was researching cyborgs and post-humanism in literature. Impressed enough by my participation in her summer course on modern American literature, she hired me to proofread the citations on her dissertation. And there, in the pages of that dissertation, I was exposed to the very grown-up, very serious world of academic debates over what constitutes a cyborg, how much human is really human, and why people are so anxious about merging with machines.

As a young man, the question struck me as essentially a philosophical one. If we're OK as a society with wearing glasses, for example, and no one thinks twice about putting contact lenses in their eyes, why does the idea of implantable brain chips evoke such strong feelings? Aren't they all just examples of correcting or augmenting our natural abilities with technology? Is it only when the technology physically enters our body that we feel it to be icky? What about pacemakers, stints and surgical screws? Why are these technologies OK in our bodies, whereas brain chips and nanobots are a frightening concept to so many?

The dissertation and my conversations with the professor that summer made me think about the issues in ways I never had, but at the end of the day it was just another academic debate, one of a million or so I was to engage in during the course of my university indoctrination. And like all those other debates, this one, too, was soon to be displaced by more pressing matters like term papers and finals.

Imagine my surprise, then, when I find the question once again muscling its way into my life as a fully-grown adult. Try as I might to escape it, there it is again: What does it mean to be human?

This time, the question sneaks up on me out of left field. After a crash course in alternative and suppressed history that led me from 9/11 Truth to the question of central banking and how money is created, I arrived at eugenics. The idea that there is a special group of people whose very genes make them fit rulers of society. The flip side of that coin is that the rest of us have genes that make us fit to be ruled, if not just eliminated from the gene pool, but the wealthy gentlemen scientists of 19th century England who first developed this theory conveniently left that part out of the PR roll for their eugenics religion. Many now know how the eugenics story goes: formulated by Francis Galton in the late 19th century, promulgated by a legion of politicians, scientists, writers and academics in the early 20th century, picked up by the Nazis (with help from generous Rockefeller Foundation grants) and forever associated with Hitler's master race philosophy and the (IBM-enabled) final solution.

What few realize and I discovered to my shock and horror was that eugenics never went away. It just went underground.

In the late 1950s the esteemed members of the American and British Eugenics Societies realized they had the mother of all PR problems on their hands. Their raison d'être was now forever associated with one of the most appalling regimes in human history. Their answer to that problem was not to abandon their philosophy as failed pseudoscience quackery, but to advocate crypto-eugenics, or eugenics by another name. Thus the American Eugenics Society merged with the (Rockefeller-founded) Population Council and began to fret about the "overpopulation problem." The *Eugenics Quarterly* journal changed its name to *Social Biology*. Julian Huxley, brother of *Brave New World* author Aldous and member of the Huxley family that had been intimately involved with eugenics since its inception, had another idea: transhumanism.

The goal of eugenics, it could be argued, was to improve the human race. And what better way to do that then by becoming something else entirely, not human, but transhuman, to transcend ourselves via technology? The idea is inherently enticing, and even more so after decades of programming through all forms of media that humanity's inevitable future is to merge more and more with this technology that we increasingly rely on.

It is now sold to us as something sexy. Our collective future is to live in gleaming plastic pre-fabricated cities, populated by hover cars and brainchipped superhumans who are capable of all the things that we have been told to dream about for years: instantaneous communication, internet access via our thoughts, nanobots in our bloodstream that will maintain our physical bodies in a state of eternal youth. Of course it's a lie, as every promised utopia in history has been, but it's a powerful one.

As an adult, this question of what it means to be human strikes me not as the rhetorical fluff of science-fiction fantasy or lofty academic debates, but an all-too-real question that increasingly demands a real-world answer. We as a society are going to have to provide an answer to that question sooner or later, or else one will be supplied for us.

Once again I turn to fiction for my understanding. This time it is Kubrick's *Spartacus*. On the surface level it's a simple story: slaves fight for their freedom against all odds. But of course it's about more than that. It's about the question of what it means to be free at all. Kubrick's answer, I suspect, is that it is not about physical freedom at all, but mental freedom. Until our minds are free, we are never truly free at all. We are never truly human until that point. That's why the ending is ultimately hopeful, despite the fact that Spartacus is captured and crucified. He may die upon that cross, but he died a free man, a human.

This is the human condition: to be born, to struggle, to learn, to grow, to fail, to die. It is a hopeful story because there is the chance to live up to our humanity, to nurture our minds, to live our lives as free men and women. We will die, but without our humanity we are never truly alive. This is what the transhumanists, posthumanists, cyborgians and their ilk wish to destroy. Our humanity, in all its glory and all its frailty.

In the pages of the book you are about to read, Aaron Franz lays bare the hidden history and true intent of those who seek to perfect humanity through technology. It is not a pleasant story, but it is one that we all need to be familiar with, especially now in the 21st century as the outrageous sci-fi fantasy of yesteryear becomes the mundane scientific reality of today.

And at the end of this journey, the question remains there for you to answer as you wish. It is a question that has pre-occupied philosophers for millennia, but it is one that we will have to provide an answer to before this century is out, or risk having one provided for us:

What does it mean to be human?

-Introduction-

Today, we are led to believe that war, cutthroat competition, and conflict are somehow going to culminate in world peace. Confusion abounds, and doublethink reigns supreme. In this world of artifice doublethink actually provides a false sense of sanity. The Modern Man has been convinced that high ideals like world peace may be achieved through warfare. There is a dialectic process at work here. It has everything to do with mind control, but it also has to do with results. Shattered minds can actually be predicted and controlled with astounding accuracy. Oddly enough, war can create peace of mind. You doubt that this is true? Take an honest look around you. You may notice people entering into impassioned states of religious ecstasy by merely chanting "Support our troops." Who, or what, are these souls paying homage to? No one answer seems to satisfy this question, but rather a complex set of spiritual revelations obtained by the honest individual are required to even begin understanding such a situation. Upon reflection the serious seeker must ask: what exactly is the religion of this world?

Manipulation of language has perpetuated war. The Word has always marked the beginning of creation. While words have power, they remain the reflection of even more powerful ideas, things that make no sound. In silence the fate of the world was conceived, and has appropriately come to life through an eternally silent population. Few are willing to stand out.

Now is the time for you to turn things around by digging deep within yourself. In doing this you will be obligated to speak out, and stand up in the face of massive resistance. This responsibility is your key to ending the Silent War. Understand that this silent and eternal war has always taken place on an invisible battlefield, the human mindscape. It is this sacred realm that has been invaded throughout time, all for the purpose of creating allegiances that would otherwise not exist. To the true seeker, the toxicity of such allegiance is plain to see. For everyone else, misery persists, but its cause remains completely unknown. Such is the war on consciousness.

Over the past century we have witnessed an escalation in the scale of warfare. As human civilization has advanced, so too has the art of destruction. It has been sufficiently proven that time and time again all sides of wars are funded by mutual financiers. International banks lend money to opposing nations. It is these globally influential private institutions which actually profit from the debilitation and weakness of all nations. The true power and sadistic wealth of this small group of profiteers is gained through the mutual weakness of their borrowers. In this depraved system the foolish nation states destroy themselves while the wise financier remains unscathed. It is not the intent of this book to detail the intricacies of this particular process, but it must be recognized if we are to truly understand how this world of ours is operating. International business is carried out in this way.

Truth is a tricky word. That is because the nature of truth lies in definitions. It is of prime importance to know that whosoever is allowed to define truth gains power over everyone else who accepts that particular definition. By agreeing upon a single version of truth, a union is formed. Consensus reality is established. Large groups of people are motivated to action according to the design of their belief systems. Proclaimed truth can be an incredibly relative thing; it isn't as absolute as many of us have come to think it is. This is an incredibly valuable fact of life.

The prevailing truth of our age is that human society has constantly improved. Throughout history various revolutions have built up to our current state. Society as a whole has benefited from this. The world we have created is a glowing beacon of progress. Alongside social growth, so too has the life of the individual become more bountiful. Technological progress has brought us comforts and luxuries that were unknown to the kings of previous centuries. Life has become longer and more abundant, and so it follows that our progressive method of social construction is effective and must continue to develop. Man must evolve further, but is it right to equate our social history with natural evolution?

Everyone seems to be in pursuit of something better and so they continue moving forward. This is usually done quickly and without question. Our methods go ignored, but our motions persist. In this way we have become automated, much like our beloved machinery. At a glance it seems as though our toil has diminished thanks to high performance tools. Looking deeper, one can see that they may have been hasty in making such assumptions.

High performance leads an athlete to victory. This is how the gold medal is won. We all have gold on our minds, but ironically we are unable to understand the profound meaning behind this symbol of wealth. Instead we see an idol in the form of a glitzy prize. Its brightness blinds us so that we cannot see its looming shadow. We

Introduction

must know that the gold of the alchemists is philosophical in nature. This is the true essence of the prize that we are unconsciously racing toward. Such gold is to be found within our selves. We never realize how valuable we are! We devalue ourselves, but we don't cease to be profitable to other entities. Fortunes have been made on the blood, sweat, and tears of the multitude. We never take the time to stop and think about what it is that we are taking part in. Our automated behavior speeds up our body while slowing down our mind, or at least its critical reasoning ability.

The actual depth of the reality that surrounds us is scarcely realized under these conditions of automation. It is rare that someone decides to slow down and reevaluate his own life. In a supposed era of increasing wisdom how could this be? It simply could not be, and we are not living in such a time. This is the Information Age, not an age of wisdom. Information need not be true or false to exist, it merely must be. We now find ourselves immersed in information. This bounty of information is all too often interpreted as wisdom, and so it is that more fuel is added to the fire of progress. The mantra becomes: if we are to become wiser, then we must increase the flow of information. Actually, an over-abundance of information leads to confusion. Confusion is not a defining characteristic of a wise individual.

Part One
What Is Going On?

-Chapter One-
The Transhuman Project

Transhumanism appears to be a new concept. It is the idea of improving humanity through the use of science. Transhumanism advocates the creation of artificial intelligences, virtual realities, life extension therapies, genetic engineering, and much more. The cybernetic merge of man and machine is seen as the way to transcend human limitations and become "better than well." The official transhuman PR campaign now falls under the banner of H+, which stands for Humanity Plus. This name alludes to all of the technological add-ons that transhumanists want to see developed to upgrade human bodies. They hope to use the decidedly human traits of logic and reason to become something better than human. Their philosophy is explained as a direct outgrowth of the Enlightenment period in European history. H+ was formerly known as the World Transhumanist Association (WTA). In 1999 they wrote a declaration of sorts, which was improved upon over the years with newer versions. Version 2.1 of The *Transhumanist FAQ* was published in 2003, and it defines their movement as,

> The intellectual and cultural movement that affirms the possibility and desirability of fundamentally improving the human condition through applied reason, especially by developing and making widely available technologies to eliminate aging and to greatly enhance human intellectual, physical, and psychological capacities.[1]

Transhumanism comes off as a completely cutting edge, high tech, and new idea to most people, and why wouldn't it? Cyborgs, implants, and AI all hold futuristic appeal. It is easy to assume that transhuman ideas are new, but such assumptions are wrong. The idea of transhumanism has been around for a very long time. The desire to transcend human existence is ancient. Of course the transhumanists point to this fact in their FAQ as well. The history of rational thought is outlined and certain mythological figures are pointed to as the philosophical champions of their ideas. Prometheus, the demigod who brought fire to men, is one of these figures. There is a monumental story lying within the thick of all this. It involves the

---

[1] *Transhumanist FAQ*, section 1.1

past, present, and future of our species. For this reason transhumanism cannot be ignored and must be understood for what it truly is.

Exponential technological growth is important to transhumanism. We are at a point in time where many different scientific disciplines exist. These separate fields of study now have greater impact on each other because developments in one area fuel further progress in another. New technologies are produced that make research easier, and this in turn makes the development of more technologies quicker. This is what defines converging technology. The convergence of all scientific achievement makes faster growth possible.

Amazing upgrades such as brain implants are spoken of by transhumanists. We are told that by putting these computer processors in our heads our intelligence will be amplified. The use of such chips has also been advocated as a way to cure certain neurological diseases such as Parkinson's. Similar implants have been developed for use in different areas of the body to enhance eyesight, hearing, and other bodily functions. Much speculation is now being done on the future of these technologies. It is possible that they could be used to widen our fields of perception and expand upon our senses. Things like infrared spectrums of light that are imperceptible to us naturally could become visible. Kevin Warwick, a Reading University (England) Professor of Cybernetics, isn't an official transhumanist, but he does go around advocating these sorts of implants. He has garnered major publicity for becoming a "cyborg." This he achieved by implanting a chip in his arm that wirelessly interfaced with a computer. By moving his arm in a certain way he could turn a light on and off. This of course was a very crude experiment that only scratched the surface of what could be done in the field.

Full integration of these technologies at a scale that would affect the lives of every living person, in fact every living thing on the face of this planet,[2] has been planned. We are slowly being braced to handle these huge developments step by step. In the 1930s people would have been disturbed to learn of Rudolf Hess's experiments in which he implanted electrodes into the brains of cats to illicit physical and emotional response. Today, Warwick's experiments spark controversy, but the long history of brain implants is rarely brought up in the process. It is long forgotten because we don't have a good

---

[2] Dvorsky's *All Together Now* claims that it is Mankind's ethical responsibility to enhance animals. This he compares to the European's "uplift" of Native Americans.

The Transhuman Project

historical memory. This is something that needs to change so that we understand the true scope of what is being laid before us under the guise of benevolence. News blurbs on Warwick have the effect of familiarizing the public mind with the general idea of implantable chips. Much publicity has been given to the idea of "chipping" humans, and there is a definite reason for this.

Very intelligent people who know when to keep quiet and when to share their work with the public carry out science. Timing is crucial and this aspect of scientific development eludes most folks' understanding. Over time standards are altered. It is through slow and steady familiarization that new ideas finally become accepted. Technology has long been building up with little to no public knowledge whatsoever. We now have to realize that a time of technological revelation is upon us. This is what transhumanism is truly about. It is here to get you up to speed on decades of secret research because the product of that research is to be used on you.

> But one must not move quicker than science, which recedes that it may advance the further   - Eliphas Levi, *The History of Magic*

Another transformational science that is advocated by transhumanists is nanotechnology. This branch of science has been a high priority project of the Department of Defense (DoD) since the early 1980s, but this should not come as a surprise. Most transhuman technology has been developed under the Defense Advanced Research Projects Agency (DARPA). Nanotech machines are actually the size of molecules. There has been success with nanotech but it remains to live up to its full futuristic potential, at least that is what we have been told. It's always difficult for a civilian to be confident in making claims about the advancement of military research. What we do know is that in theory nanobots could be programmed as assemblers. These nano-assemblers could build up materials atom by atom. They could make any chemically stable molecule by arranging atoms precisely. This would allow the creation of anything imaginable. Nanotech could take the atoms from an otherwise useless source and turn it into something useful. You could turn dirt directly into food with nanotech. The possibilities for this technology are seemingly endless. Molecular manufacturing is hailed by transhumanists as a way to conquer scarcity. In a scarcity-free world people would be able to concentrate on things other than survival. They would be able to develop higher pursuits that enrich the mind instead of being condemned to a life of misery. Nanotech has the ability to change everything. Its size would allow it to flow through biological bodies and alter them from the inside out. This of course holds much transformational potential and nanotech is perhaps one of the most important technologies to develop successfully if the ultimate transhuman dream is to be realized.

The ability to repair and maintain our bodies at a molecular scale could enable incredibly effective life extension therapies. Gerontologists such as Aubrey de Grey are already saying that it will soon be possible to extend healthy human life indefinitely, and this is without the use of nanotech. The opportunity to essentially live forever is now within reach. This is causing many interested people to extend their natural life span as far as possible with current methods in the hope that they will live long enough to be able to receive advanced forms of life extension in the near future. Today, caloric restriction, dietary supplements, and healthy living are the preferred methods used by those who hope to make it to de Grey's theoretical point of "escape velocity." If you can receive dramatic life extension soon enough, the technology will continue improving to the point that you may literally live forever. This prospect is so appealing that some people are choosing to freeze their bodies immediately upon their own death. Cryonics companies have been offering this service for decades now. Ever since the inception of the cryonics movement in the late 1960s, its adherents have been concerned with the many transhuman prospects outlined here. Their hope lies in preserving the body just after clinical death so that it may be brought back to life in a technologically advanced future. Some people have their whole bodies frozen but many opt for the cheaper and presumably more effective cryonics procedure. They undergo postmortem surgical decapitation and their head is frozen. The idea behind this is that in the future all you will need to exist eternally is a brain.

The concept of mind uploading has to do with copying a person's brain onto a computer. A powerful enough system may be able to transfer your very thoughts into its own digital circuitry and effectively reproduce your personality. By sending nanobots into your head to record the structure of your brain, you could presumably copy it completely. There is no way of effectively doing this now, but the uploading idea centers on life in a computer-generated world. The computer that houses your mind could also run a program that simulates the physical world. A better version of reality could be programmed for you to live within. The possibilities would be endless and in this way the concept of uploading ties in directly with the creation of advanced virtual reality. Life could be lived entirely within computers if uploading somehow became possible.

The Transhuman Project

Transferring your brain to a computer could also allow the control of a non-organic body. Your brain could become the pilot of a robot. This concept gets really strange really quick, and makes for fun speculative rants on transhumanist forums. Speculation is hard to resist and we must work hard to remain focused on what is actually happening in today's world. For now uploading remains science fiction, but the possibilities for advanced virtual reality do seem within practical reach. Beyond this, data collection and pattern recognition systems that scour Internet are already being used to run relatively primitive simulations for the purposes of prediction. Such "agent-based societal simulations" have become high budget military projects. The Pentagon has just such a program in operation; its Sentient World Simulation (SWS), which is used to simulate disaster scenarios. In turn SWS is used to help the military craft and test PSY-OPS in the context of these scenarios.

Advanced robotics and brain-machine interfaces (BMIs) are very important technologies that relate to transhumanism. There have been many university and military projects to create robotic exoskeletons that enhance human strength. These projects have been co-ops between the military and academia, and you will find that DARPA funds and oversees most of this university research. The robotic exoskeletons resemble the familiar mechanical suits of science fiction movies and video games. A person stands inside such mechanical skeletons and controls them with their own body motions. These sorts of technologies have been surprisingly successful, and have been around for a number of years. Brain-machine interface would allow the advanced control of these types of machines and much more. Through the use of special helmets, people's brain activity can be read. When this information is run through a computer that understands what specific brain patterns mean, this information can in turn be used to control a machine. The end result is that you think about making a robotic vehicle go forward and it actually begins to move forward. Nearly all transhuman technologies have been developed by the military. This is nothing new for technological development and should come as no surprise, but it is a fact that often gets lost in that magical first moment in which the public sets their eyes upon the latest gizmo. We are beginning to see an explosion of these technologies on the consumer market. Brain machine interfaces are popping up in the video game industry and other forms of entertainment. Their first purpose is to amuse us, but soon afterward they will be used for serious applications. In fact, that is what they were created for from the start, but we were never meant to notice this. Our purpose is to produce, consume, and transform, not to think.

Mind and machine are blending, and as they do a strange new amalgamation is appearing. We see this with brain-machine interfaces. In their case, mind is still controlling matter, but in other converging technologies we are beginning to see mind arising from otherwise lifeless matter. The creation of artificial intelligence is groundbreaking because it has to do with the actual production of mind. The abstract concept of mind itself is becoming tangible. It is being reduced mathematically to patterns and algorithms. By setting up the right startup code, a thinking machine could be created. But how well will these artificial intelligences be able to think? How far will AI technology go? This is hard to say. Transhumanists believe that an Artificial General Intelligence (AGI) is definitely possible, and that it is only a matter of time until one emerges. Leading transhumanist Ben Goertzel is confident of this, and his opinion should be respected, as he is also one of the most prominent AGI researchers. As he puts it, AGI would be marked by its ability to "perform complex patterns in complex environments." In other words, it would be able to adapt and learn. It could grow to be as intelligent, or more intelligent, than we are. This is the sincere hope of Goertzel and other top AGI researchers and developers. They believe that when AGI becomes powerful enough, it will be able to lead our world. It will take over the task of scientific development, and take it to an astonishingly fast pace.

The Singularity

Before we begin it is important to understand what is meant when people use the term singularity in conjunction with technology. The technological Singularity is now an established idea with its own identity, but the word *singularity* itself is not new. In science the term singularity is used to describe a point at which the laws of physics fall apart. The inability to comprehend certain cosmic events is actually one of the main motivations behind the creation of the technological Singularity. An intense drive to know fuels the mission to enhance intelligence. We want to know how the universe was created, but this is not yet possible. We want to unravel the mysteries of existence. Only an intellect of extreme magnitude could do such a thing. This is why the Singularity revolves around the creation of an artificial superintelligence.

The Singularity is an important event in bringing all converging technologies to fruition. This hypothetical point was first posited by Vernor Vinge, and has since been promoted by many of the same

The Transhuman Project

people who head up the transhumanist movement. The idea is to truly converge all separate technologies into one singular force, which will become far more powerful than any single piece of technology ever could be on its own. It is already apparent today that we are at a high level of technological achievement, but not quite high enough to achieve a complete transhuman merge of man and machine. To pull off such a feat would require the intervention of some sort of superintelligence. It is this need for divine intervention that really defines the Singularity. The Singularity scenario is built upon the premise that Artificial General Intelligence will eventually reach a human-level thought process. It will be capable of logical and independent thought equal to that of an intelligent human being. Once this happens AGI will begin to improve by developing new cognitive upgrades and applying them to its own hardware. It will become increasingly intelligent until it eventually reaches a level of superintelligence. This simply means that it will be so far beyond our mental limits that it will be imperceptibly intelligent from our human point of view. As AGI self-improves, it will also be conducting scientific research at an incredible pace and, as a result, new technologies will be developed for all purposes. It could spawn other AI's that could be put to the task of aiding its own growth or performing other duties that need to be done to further a broad range of goals. It is difficult to impossible to say what these goals would ultimately be.

The Singularity scenario would create an exponential increase of all scientific knowledge and it would change the entire world in the process. Development could speed up to such a pace that not even humans (the most adaptable species on this planet) would be able to adapt to the constant changes incurred. Average human beings would be left behind in the wake of massive change with no way to keep up with the pace set by AGI. What on earth could humans possibly do in such a situation? This question favors the transhumanist movement. It is their cue to tell us that by merging humanity with technology, AGI could actually save us from certain extinction. We could supposedly keep up with the Singularity by enhancing our own brains with new technologies created by AGI. It has been suggested that we would even be able to merge with AGI itself in order to unite with its superintelligence. This concept is similar to that of a global hive mind, a thing to be achieved by linking individual human minds together through some form of technological network in order to build one overarching and collectivized brain. We can see in these examples that the creation of superintelligence is often linked to the idea of collective unity. The joining of separate individual intelligences in order to expand the power of the One is most definitely a powerful philosophical influence that reveals itself in many transhuman concepts. The very term Singularity refers to this notion; that of one, singular, thing. Through Singularity separate technologies, sciences, and human minds come together to create one massively powerful over-mind. This is done to improve the power of the collective.

The Singularity ties in with the concept of the Omega Point. This is yet another theoretical form of convergence in which all life culminates into one large super organism. The idea of the Omega Point is born out of old philosophies such as panpsychism, which itself is the belief in a collective consciousness. Panpsychism posits that all matter possesses some degree of consciousness, and taken as a whole, the entirety of material creation is viewed as one large being. This being has been broken into a seemingly infinite amount of pieces and is now dancing a cosmic ballet of existential convergence. An unfathomable amount of apparently separate objects, things such as galaxies, stars, planets, minerals, plants, and animals are all part of this larger being. On their own they can only reach a certain degree of consciousness, but as a whole they create something of greater value. Higher forms of material life have achieved a degree of consciousness, which allows them to recognize their place in this grand scheme of things, and thus they willingly attempt to accelerate the grand purpose of existence itself. A human being who has come to uphold the Cosmist philosophy (a transhuman philosophy that relates to panpsychism) is a prime example. Upon divine revelation, they decide to make it their goal to unite the disparate pieces of the entire universe. By imbuing intelligence into matter the microcosm becomes self-aware, and thus wakes up to the fact that it is a part of the larger intelligent macrocosm. This sleeping giant, the entire universe, or Primordial Adam of the Cabalistic tradition, is itself awakened at the Omega Point. The transhumanist created philosophy/ religion of Cosmism has much in common with Cabala and other mystic practices.

There has been much writing done about the Omega Point by the Jesuit priest/ paleontologist/ philosopher, Pierre Teilhard de Chardin. He was a man credited by many current transhumanist thinkers with the creation of their own current philosophies. He is a prime example of the fact that science, religion, and philosophy have never been separate. The only separation between these disciplines has existed in the minds of shortsighted human beings. This separation has indeed been packaged for mass consumption in modern times so that the enlightened ones, those who know better, may work on their heavenly plan without having to hear the nagging voices of dissent from a populace not in line with their personal views of the universe. The cosmist viewpoint aligns with the ancient Egyptian philosophy of the universe, which says that everything in existence is all part of a single universal law. Now that the dawn of a new age is upon us, we are all being raised up to an intellectual level in which this unified view returns. A "New Renaissance" is emerging, and is giving rise to a philosophically, spiritually, and technologically driven science. Implementing this new form of science involves telling humans that they no longer have to remain mortal, that they may live eternally in the style of a god. This may sound like an exaggeration, but one has to seriously consider the claims of gerontologists, that human life may soon be extended indefinitely. Combining life extension with "enhancement" we can see an eternal promise taking on an all-new form before our very eyes. Tranhumanists believe (whether consciously or unconsciously) that the average person will soon be lifted to a state of relative godhood. They are convinced that this is going to happen, and they do use the term *god* to describe it. Before getting too excited about future possibilities we must look at our present and past reality. Has the average person ever really received something for nothing? What will the price of immortality be, and will it be something far beyond the bounds of monetary concern?

Our place in time has to be considered humbly, and this can be hard to do when faced with technologies that promise so much. Please be aware that throughout thousands of years of human history the common people have consistently been fed half-truths and outright lies. The Singularity is being billed as the way to literally fulfill the religious promise of the ages; that of second life. The masses throughout the ages may have believed in false gods, but now we are led to believe that our new god (the god of scientific logic and reason) is infallible. The ability to transcend our human existence now appears to be physically possible. The ramifications of this are staggering. Our world is now going through tremendous changes in a last push to reach the fabled Omega Point of cosmic singularity. The ends justify the means for those inclined to believe in the virtues of this project. Individual human lives have little value when compared to that of the universal consciousness. Mass sacrifices are acceptable to those in pursuit of the greater good. This is the mindset of those who are fully committed to bringing about a singularity. It is important to point out that this entire philosophy is based on speculation, and in truth, we do not know the full ramifications of what will come of all this. What we can see clearly is that the

Singularity has everything to do with the grand quest of becoming one with everything that exists.

We now bear witness to the possible end of the human era in favor of something better, something that would embody the creative height of Man's achievement. This new creation, being molded and programmed in our image, could accurately be labeled the Son of Man.[3] This son will be the heir to Man's throne, and take technological wonders to make them his own. In a way, you could compare Singularity to a scientific day of reckoning. The technological apocalypse, as some have called it, could bring about a radical new dynamic within the universe. Whether or not the original creator, Man himself, would be around to see his son's work is impossible to say. We simply don't know what we are getting into, but that doesn't stop some from acting as if they do know. Many Cosmists who do at least realize that Singularity speculation is pointless, simply do not care that humanity may eventually be taken out. They have *trans*cended care, by revering ideals that far outreach the bounds of mortal man.

What Cosmists know for sure is that any singularity scenario brought about by AGI superintelligence would be preferable to this dull human state of affairs. They truly don't mind seeing humanity obliterated. Be aware that some highly intelligent people feel this way; people with IQs far above your own. They are committed to their transcendent belief system.

Science is a collective process, and accordingly the consciousness of scientists themselves is collectivized. The apparently conscious decision which drives a particular person to be a scientist leads them to build upon a foundation which was laid by many others who came before. By adding to this collective pool of human achievement they get the satisfaction of being part of something larger than themselves. This yearning has sparked *Fire in the Minds of Men*, and has led them to dedicate their entire lives to the hope of building a better world and to expand the knowledge of the human race itself. This is not necessarily a bad thing but an interesting point arises when contemplating this situation. The individual human is putting his effort into something that he could never completely understand.

---

[3] Religions have already begun to buddy up with transhumanism. The Mormon Transhumanist Association believes that converging technology represents the practical means by which their Fourth Epoch/ Millenium and Transfiguration may take place.

The Transhuman Project

The whole of scientific knowledge is often viewed as an entity in its own right. This idea is seen in Ben Goertzel's *Cosmist Manifesto*. He explains that the mind of science has its own identity. Oftentimes a person works his entire lifetime on building up this collective mind without ever stopping to contemplate his own. He in turn identifies more with the collective mind of science than he does with his own personal self. It is in this way that a sort of sacrifice occurs. The individual human consciousness is sacrificed to the greater consciousness of the macrocosmic sleeping giant described by panpsychism. The great irony here is that there is actually a lack of conscious thought on the part of the scientist. Conscious thought is separate from unconscious action in the fact that it is self-aware. To be truly conscious of one's self requires recognizing unconscious actions and drives. When one is unaware of his own motivations, or does not take the time to understand his own mind, how can he truly be deemed conscious in the first place? It is very telling that the mind of science is built upon this foundation, although many would be quick to refute this point. They would say that scientists know full well what they are doing and why they choose their line of work in the first place. They know from the beginning that they are aiding the progress of mankind as a whole. Maybe so, but do they really know what the Singularity is? Is it possible that they ever could?

It must be noted again that we simply do not know the full reality of the Singularity. It is very possible that no such event will ever even happen. The arrival of a superintelligent artificial mind may never come. The Singularity may very well be a convenient distraction from the reality of converging technologies today. Yes, technologies are converging, but this does not mean that AGI will come about and create a complete singularity. AGI itself may be overemphasized as a dream that may never be realized. It may never truly exist, and even if a superintelligent mind came into being, we may never truly know if it is aware of itself, or if it is programmed to act as if it has conscious self awareness. Its consciousness could be genuine but so far removed from our own that it would be completely alien to us. All these possibilities remain while the whole truth remains unknown.

What we do know is that narrow AI systems do exist, and at present are being used for many purposes. AI has been put to the task of mining patterns within masses of information gathered during scientific research. Artificial intelligences are finding patterns in biological and social databanks. A detailed description of human genetics and psychological motivations is being written with the help of these machines. They are being used to understand every last detail of human biology and behavior. This is seen in the bioinformatics research of such agencies as the Center for Disease Control (CDC), and Department of Commerce (DOC). We also see this trend in the development of massive AI surveillance systems such as Project INDECT of the European Union. The artificial intelligence machines themselves may not yet be conscious but they are serving as barometers for the collective human consciousness (or perhaps more appropriately, unconsciousness). Such tools can be exploited by anyone who wishes to develop a world society according to their own private interest. A major trend in AI is to better understand humanity so that more efficient directive influence may be employed en masse. This is a simple issue of management. The scale of human management continues to increase with the power of technology. This cannot be denied. Is the publicly promoted idea of a singularity simply a diversion away from the real power and direction of converging technologies? This is a possibility that needs to be considered when we understand the true state of affairs within our present reality.

The term singularity is certainly appropriate within the context of a unified world system. Such a world union would combine social, political, and economic systems to create an overarching command and control singularity. It wouldn't take an AGI to run such an operation, but narrow AI systems like the ones that already exist could certainly be used to intelligently manage the entire world. This appears to be the way that things are going.

The destruction of all things private: property, identity, and family, is being carried out methodically. This trend may culminate in the birth of a new world system, a singular behemoth that takes all aspects of human life and reduces them to digital nodes. These nodes would be organized in the most efficient way possible in order that the singularity, or world society, may develop further. The directive influence of government in the personal lives of individuals continues to grow on a daily basis. How could these current trends lead to a genuine world peace? When does the dissolution of individual human identity ever end? How far could we travel down this road without noticing the direction we are headed? These are incredibly important questions that need to be explored by every human being on this planet. Entire books have been dedicated to the Singularity. This is understandable but sadly these books are very much lacking in their criticisms of the very idea of Singularity itself. Instead they focus on how wonderful the future may be thanks to artificial intelligence and all of the transhuman enhancements that it shall spawn. It is easy to speculate on the future of technology because we are now at a point in which anything is possible. However, before we get carried away, we must first be aware of the current state of affairs in this world. We have to see the trends that are present now; these are continually pushed in a certain direction. It must be understood that there are directive forces in this world. These forces manifest in many forms such as governments, nongovernmental organizations, corporations, individuals, and so on. The very idea of technological Singularity comes from these sources. So, what exactly is the big idea behind the Singularity?

Alchemy and Transhumanism

Things are not always as they seem. Appearances are, more often than not, superficial masks made to disguise the full truth. This applies to everything in our world. There is no exception for the supposedly clear-cut and objective sciences. Because of our surface level assumptions, the history of science itself is misunderstood. We believe in a generic idea of science without ever pondering its true history or meaning. We don't think of science from a philosophical point of view. This is because we tend to divvy up our thinking into categories. We think of science, philosophy, and religion as separate disciplines. In truth, these are all segments of one larger whole, one big idea. They always have been. So what exactly is the big idea? The idea is a unified cosmic view of the universe. Physical laws are the business of science, high ideals are that of philosophy, and the unknown is recognized and made tangible through theology.

A synergetic view of the cosmos has existed at least since the birth of history. Ancient Egypt recognized the unknown. Its unified understanding of the universe included early science. Mystic philosophy was held sacred and filled with secrets that were only known to a select few. Alchemy itself is known as **the Royal Secret**,

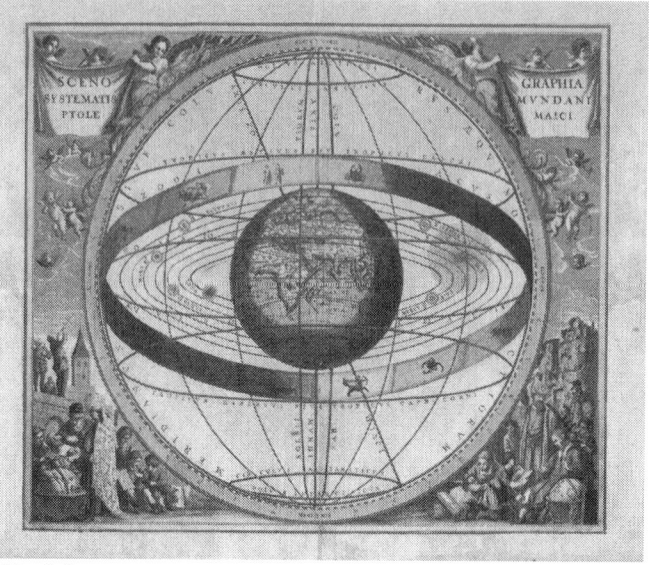

Fig. 1.1 Ptolemy's universe. Angels fly high above as men with compasses labor down below. Astronomy has always intertwined with Myth. The zodiac is important within both astronomy and astrology (science and religion)

as it was derived from these origins. Egypt is referred to as the black land, or Khem, and the Black Art of Al-Khem-y was used to control its population. Black refers to the darkness of ignorance that enslaved the masses and enabled the few who could see to rule over them. Secrets were hidden from mortals who had not gone through the divine rites of the priesthood.[4] It was this priesthood who held the true reins of power in the black land of Egypt, as they were the ones who understood the social sciences, which have always been the key to controlling populations. The scientific understanding of human behavior has been exploited ever since humanity was civilized. Civilization is indeed a construct of the mind, and it takes the repetition of ideas to maintain it.

A mind divided is a mind incapable of logical thinking, or profound self-reflection. Ideas that are established as incompatible are forbidden from mingling amongst each other. The establishment of

---

[4] It is well known that pharaohs were worshipped as living gods. This high status was delivered via the priesthood. A self-proclaimed divine government was established in this way.

The Transhuman Project

inner conflict and contradiction within the minds of men has always debilitated them. When thoughts are allowed the freedom to work together then they grow into larger and more complete ideas. Thus science, religion, and philosophy were ultimately to be separated so that the masses would be left in metaphorical darkness. The scientific application of mind control has been around for a very long time. The collective human psyche has been smashed into fragments, and now seems unable to become whole again. This is what we long for. The search for ourselves is a never-ending battle against those who feed us lies and half-truths in order that we remain broken.

No being in touch with his inner and outer nature could ever fall under another's control. Today however, our minds have become so weak that simply disagreeing with established paradigms causes us pain and suffering. This occurs in more ways than one, as we both feel shame for simply contradicting any norm, and we cringe at the threat of physical pain that arises when confronting the establishment itself. The truth hurts.

There is a synergy of science and the ancient mystical practice of alchemy. Behind the transhumanist movement are some very ancient and profound ideas. Not least of which is the hope of achieving the Elixir of Life. The Philosopher's Stone is another name for this holy grail of immortality. An insatiable yearning for supreme knowledge has forever existed and has taken many forms throughout the ages. It can now be seen throughout science. Unknown to casual observers and oblivious technicians, the sciences are working toward an ultimate union of mind and matter.

Transhumanism is by no means a new idea. It is simply a new interpretation of a timeless philosophy as well as a rational plan for its incarnation within the physical world. The ancient mystery tradition promotes the concept that all matter within the physical universe is in various stages of *becoming*. This means that everything within creation is on its way toward becoming or merging with the unseen creator itself. From this point of view it is actually possible for material bodies to become divine. Man may become and/or merge with God itself. This belief drives the transhumanist philosophy, which some transhumanists/ AI enthusiasts have called Cosmism. The goal of the Cosmist is to reach out toward the greater collective intelligence, which they believe exists throughout the Universe, so that they may literally become it. It is to expand the power of mind until, philosophically speaking, it reaches a level of godhood. This is why transhumanists frequently speak of "becoming gods" and they certainly aren't joking around when they say this. They truly mean what they say; make no mistake about that.

The concept of becoming was summed up well by nuclear physicist and human cloning advocate Richard Seed in an interview he did with National Public Radio in 1997. During this interview he said,

> God made man in his own image. God intended for man to become one with God. Cloning is the first serious step in becoming one with God. We are going to become one with God.

There you have it. This man believes that it is not only his right to become one with god, but that it will indeed happen. This spiritual ascent to godhood would be complemented by the physical attainment of that high ideal. Scientific knowledge not only contains the key to gaining god-like power, but rather to becoming God itself. Seed later had this to say in the excellent film *Technocalyps*,

> We are going to become gods, period. If you don't like it, get off. You don't have to contribute, you don't have to participate, but if you are going to interfere with me becoming God... gonna have big trouble. We will have warfare.

It's easy to see how this philosophy will clash with predominating world-views. Seed's mention of warfare hints at future conflict that will arise when opposed ideologies crash into each other head on. This is what may happen as technology becomes increasingly powerful. People who are scientifically minded will clash with those who come from a religious point of view. This conflict has actually been going on for some time now and will surely increase with the level of technological development. The population has been divided (in their thinking) so that they waste their time bickering amongst themselves as real change continues all around them.

The name Richard Seed is itself incredibly interesting as it refers to the male reproductive system. To those who don't quite follow me, just think of the shorthand version of the name Richard. Got it? Ok, let us continue. The word seed is of course symbolic of the spermatozoa, which is one of the two players that come together to regenerate the human race at the microscopic level. His name is a sexual reference. It alludes to the implication of his life's work. By modifying the act of reproduction with science, man may recreate the act of creation itself. This is the occult significance of human cloning. It enables man to become a god. Man for the first time may take his reproduction away from chance, and steer it wherever he so chooses. Under this new method of creation every being could be designed from the genetic level up. Combining genetic engineering with the use of artificial wombs one could supposedly create a new and better race. At such a point, man would become the godlike giver of life. He could choose to give life to a more perfected being of his own design. This is how man achieves godhood through his knowledge of the physical world.

The World Transhumanist Association, which has since changed its name to H+, has published a document outlining its basic goals, philosophy, and history. The *Transhumanist FAQ* is a very revealing read for anyone who is aware of the true nature of science. It claims that a posthuman future won't necessarily be the end of the human race; it will merely be an improvement. In a section that outlines the philosophical precursors to transhumanism, alchemy and magic are mentioned specifically. Wise transhumanists are well aware of the true meaning of alchemy. They know that their own seemingly new ideas are simply evolved versions of the Black Art. Unfortunately for us, most of the public remains in metaphorical darkness, ignorant of the fact that transhumanism even exists let alone that it is a new incarnation of alchemy. Even among its own followers, the connection to alchemy is not well understood. Many transhumanist foot soldiers are in fact atheists. These are the transhuman fundamentalists who are unwilling to consider reality beyond the scope of their own physical existence. They are locked into the science fragment of the divided brain, unable to broaden their horizons due to mental dams they have constructed within their own minds. They have made up their own minds, just as religious fundamentalists have. This makes them perfect followers for a priesthood to control.

The fact that alchemy and magic are indeed related to transhumanism is freely admitted by the WTA. In the *Philosophical and Cultural Antecedents* (section 5.1) of their FAQ they say,

> That people really strove to live longer and richer lives can also be seen in the development of systems of magic and alchemy; lacking scientific means of producing an elixir of life, one resorted to magical means.

In this quote we get a glimpse of the practical element of magic. All magical practices are done to effect physical change in this world. The actual significance of this fact will remain elusive, and this is appropriate, because so much magic relies on illusion. In stage magic the audience is never supposed to understand how the magician performs his trick. There may be a perfectly reasonable and scientific explanation for it, but the sheer spectacle of the show is designed to distract people from what is truly going on. Sleight of hand tricks fool the eye. They misdirect attention away from where the explanation lies.

The magician, being a craftsman of illusions, restructures the material universe in a way that appears supernatural to the untrained observer. In technology the same rules apply as technicians take on the qualities of wizards and sorcerers. They learn how to alter matter beyond the untrained eye's ability to see and the unenlightened mind's to comprehend. This is how people are *blinded by the light*. If you cannot see that technology is acting on the material universe, your mind will revert to a simple and incomplete explanation. You will believe in magic. Those who lack knowledge of the higher sciences will have to accept the sorcerer's dubious explanation. Beyond merely impressing people with a good magic show, technology has the ability to dramatically reshape the world according to the desires of those that control it.

In section 5.1 of the Transhumanist FAQ it states,

> There is a tendency in at least some individuals always to try to find a way around every limitation and obstacle.

These individuals are the alchemists who have set themselves to the task of perfection. They do this by using every tool they have at their disposal.

Before being able to manipulate the material world, the alchemist must first perfect his own psyche. The transformation of mind is the first step taken on the road to becoming an alchemist. Alchemy concerns itself with the attainment of the Philosopher's Stone. Given the fact that this particular stone is philosophical in nature, it follows that it is made of ideas. All those who believe in the literal translation of "turning the base metals into gold" have been fooled. All of the terms used within alchemy are simply metaphors. Alchemical literature is allegorical. Its purpose is to keep its power a secret to those who are deemed unworthy. The Philosopher's Stone is often symbolized by diamond. The untrained observer would take this literally and consequently miss the point. The diamond represents the perfected man, one who has realized his own divine potential. The perfected stone is the perfected man. It is made by taking the raw material, the rough ashlar, the ignorant neophyte, and putting him through a ritual initiation that shows him the Light. Diamonds are cut in such a way that they reflect light intensely. Enlightenment is synonymous with knowing. Awakening the intelligence inherent within the mind of man is akin to the cutting of a diamond. Intelligent people are referred to as being bright. The ancient mystery tradition

The Transhuman Project

has always involved the recruiting of the best and brightest minds available.

Before its modern incarnation, academia was limited to a small portion of the population whose minds were raised up through ritual initiation. Such rituals were intimately linked with practical scientific study. There were the apprentices and their knowledgeable masters. To this day this design can be seen in the blue degrees of Freemasonry, with the first degree being the Entered Apprentice and the third being the Master Mason. The enlightened ones take oaths of secrecy and allegiance to one another before they are taught how to manipulate matter. They learn how to make fortunes by taking base metals and turning them into gold. There are many ways to interpret this allegory, but in the end real wealth and influence are indeed accumulated by these alchemists. Who exactly is a member of a particular fraternal order or secret sect is difficult to say, but the influence of the ancient Mystery tradition on our modern world is well documented. The truth has been told, but many have not listened. Cryptic symbolism is the preferred language of the alchemists. Because of this the majority of people who are still in the dark can't hear the message.

There are many interesting cryptic messages to be found in scientific literature. I remember reading a quote about nanotech that claimed it could help turn coal into diamond. This is a direct reference to alchemy, as turning coal into diamond is one of the most used metaphors for transmutation. Alchemists are sometimes referred to as miners, and the process of turning carbon into diamond is akin to the philosophic act of perfecting the stone. Alchemists find promising young minds (apprentices otherwise known as philosophical sons) and improve them so that they may shine all the brighter. Transmutation is an interesting term in and of itself. It references a chemical change that allows the alchemist to perfect matter. A mutation is a rapid transformation of one substance into another. Through transmutation the alchemist causes change to happen in the physical world at a very rapid pace. This is an important aspect of the art of alchemy. It seeks to cause rapid change, to go beyond the bounds of nature by forcing the raw Prima Materia to mutate quickly according to will. Alchemy is a way of speeding up the evolutionary process. For the alchemist Nature is just too slow to bear. Beyond this, the fact that random genetic changes take place is also viewed as a problem. Through applied science, man should take hold of chaotic nature and direct evolution according to a well-laid plan, instead of allowing random chaotic change to occur. The will of man is seen as the ultimate guiding force on this planet, and as such the complete transmutation of the natural world is the essence of the alchemical Great Work. From this philosophical viewpoint it is man's destiny to reshape the natural world into something better. Complete and total change is desirable if it is done intelligently through logic and reason. To literally alter the natural world is the highest esoteric goal of alchemy and it brings the ancient art full circle from allegorical self-perfection to actual physical transmutation. Alchemists just love circles.

Now what is transhumanism about, if not the perfection of the natural world? The transhuman goal is to use applied logic and reason to upgrade human biology, to transmute the human body to a new and better posthuman design. It relies on the human intellect, a powerful engine of creation. Transhumanists don't just want to better humanity; they also speak of uplifting animals through technological enhancements. Their ideals reach far beyond the realm of earthly existence too, as they frequently speak of traveling throughout space to harness the wasted energy of stars so that it may be used to create something useful. All things in their natural state are simply not good enough. Nature is inefficient, violent, and chaotic. This is to be changed according to the will of god. The will of god within this philosophical method is the will of man; it always has been. This too is an esoteric secret that has been confined to the priesthood throughout the ages. In transhumanism we see a public display of this ancient philosophy. This is very important to realize. Transhumanism presents a direct message to all who can comprehend its enormous implications. We live in an extremely important point in time. Human power is culminating in converging technology, and there are definite plans to use this power to alter life as we know it. We are all being led unwittingly to the altar (the place of mystic transformation). This is why transhumanists speak of building and becoming gods.

Transmutation is a powerful word. Transhumanism shares the same prefix, *trans*, which implies movement. Going from one state to another is inherent in all words containing this prefix. Transport, transition, and transcend are all examples of this. Movement from one point to another or a panorama of some type is inherent in words with the *trans* prefix. To transmute is to transform, to change from one chemical state to the next. The metaphorical chemistry of alchemy can be used to symbolize a great many things. The transmutation of the mind itself enables intelligent transformation of the physical world. There is a reason as to why the term transhumanism was adopted by people who want to see a transformation of the human

The Transhuman Project

race through the use of technology. The three-fold transformation proposed by transhumanists is this: 1. Human; 2. Transhuman; 3. Posthuman.

1. The human state: could be labeled the Prima Materia, for it is the primal base substance upon which an improved creation will emerge. Imagine the human body as a stone. By shaping this stone in the fashion of a master sculptor, the art of alchemy is realized. As a quick aside, many of these terms: stone, sculptor, and art, are very powerful alchemical metaphors. They find their way into transhuman and scientific literature very often. If you have eyes to see, you can't miss them.

2. Transhuman: the second phase is marked by transformation itself. The transmutation that changes the human into a transhuman is a phase of movement. The transhuman phase is not permanent. It is merely a transition. To completely transform, transmute, and transcend the limits of human biology is the ultimate goal. After the artist has perfected this stone, an ideal state of being is achieved in the final step of the process, the New Man (of the mystic tradition) by another name.

3. Posthuman: It is unknown what this life form would ultimately be, or if the transformational process would ever even stop, but the theoretical end product of scientific transmutation is the posthuman. This creation would mark the end of the human era, and usher in a whole new form of existence. The transhumanists believe that this would be a dramatic improvement over naturally evolved humanity. The ultimate goal is to infuse intelligence into matter, to transcend the limitations of the cold natural world, and rise above.

For those who are aware of the Mystery tradition and of ritual initiation, it is clear to see that this threefold process actually represents the death and rebirth ritual itself. Some serious TransAlchemy is afoot. Now let us discover its cause.

-Chapter Two-
It's Official

> The ability to see complex systems at the molecular and atomic level will bring a New Renaissance... The collective multiphenomena and multiscale behavior of systems between single atoms and bulk become the center of attention in order to extract new properties, phenomena, and function - like a new alchemy.
>
> -National Science Foundation and Department of Commerce sponsored event *Converging Technologies for Improving Human Performance*, Dec. 2001, page 80 of the report[5]

While most people remain in the dark, completely oblivious to the true potential of converging technologies, there is a small portion of the population that understands the power that science currently holds over society. This power has been well known for a very long time, but now rapid development is spurring on massive technological expansion. Converging technologies have unlimited potential. Their creation would undoubtedly alter the world dramatically. This is in fact a high altar of spiritual importance and some of the most serious scientists realize this fact. The future is normally perceived as a distant event but given the current speed of progress the future is no longer tomorrow: it is today. A constant developmental process happens to coincide with our everyday lives and most of us remain oblivious.

The possibility of a technological Singularity has been given serious consideration by high-ranking individuals, institutions, and governments. These entities are not foolish enough to ignore the potential power inherent in further scientific development. Who do you think has been developing the technology that has led up to this point? Behind powerful technologies lie powerful people and meticulous planning. An integrated system of government, military, academia, NGOs, and corporations exists as the creative force behind emerging technologies. The fact that much planning has taken place to ensure a profitable Singularity has not been a focus of media attention. We, the people, aren't supposed to realize this. The closest explanation given in terms of research and development has come in

---

[5] Conference held in Dec. 2001, report is dated June, 2002.

the form of short news clips, each outlining the latest scientific breakthrough. Ray Kurzweil has addressed the Singularity and related topics in his various talk show appearances, every time plugging his latest book. It is in this way that people are given a general idea of what the Singularity is, but simply reading one of Kurzweil's books is not enough to get an understanding of what this issue is all about and who is behind it. To gain real insight one should look to the many official whitepapers which have been written up by governmental institutions, NGOs, private corporations, and think tanks. Large conferences have been held regarding converging technologies and the direction that they should be steered.

It was early December in the year 2001. Less than two months after the infamous 9/11 terror attacks, while most people were still in a state of shock, shaken from the violence that befell our nation, the United States Department of Commerce (DOC) and the National Science Foundation (NSF) sponsored a gathering of awe-inspiring magnitude. It was a conference which brought together experts from academia, private firms, and government to talk about *Converging Technologies for Improving Human Performance: Nanotechnology, Biotechnology, Information Technology and Cognitive Science* or NBIC for short. These four specific areas of science were chosen as the most important fields of research. The convergence of NBIC was explained as the best way to achieve a technological Singularity, and more importantly achieve the high goal of "improving human performance."

The transhuman concept of enhancement was explained in great detail although it was never specifically labeled as transhumanism. Enhancing individuals would subsequently streamline the entirety of civilization to create "a more efficient societal structure." A better world could be engineered from the ground up. The current human form could be genetically altered, or at the very least enhanced through greater connectivity to hyper-efficient machines. Interconnectivity between everything and everyone was seen as the ultimate form of pooling resources to achieve answers to many of the world's problems. According to this group, "a New Renaissance" could emerge. This was explained as the ultimate way to integrate humanity with nature and establish an entirely new sort of world. Optimism was on full tilt as evidenced by quotes such as this,

> Moving forward simultaneously along many of these paths could achieve a golden age... The twenty-first century could end in world peace, universal prosperity, and evolution to a higher level of compassion and accomplishment... It is hard to find the right metaphor

to see a century into the future, but it may be that humanity would become like a single, distributed and interconnected 'brain.'[6]

Not only is convergence focused on bringing technologies together, it is fundamentally concerned with amalgamating the entire human race. Our cultures, societies, and minds are to be consolidated due to these coming changes. Singularity is all-inclusive as it seeks to merge everything and everyone. The NBIC report lays out many ways in which this may be achieved. It predicts the advancement of communications technologies and the corresponding effect they will have on human society at a global level. Communications will fundamentally reshape our relationships and interactions with the world around us. All communications technologies that existed in the year 2001 would eventually merge with virtual environments. As humans continue spending more time in virtual realities, they will also be using more and more personal devices that allow them to gather data on the real world around them. This represents a convergence of real and artificial, labeled "augmented reality."

Due to increased virtualization and use of digital devices, information will be constantly accessible at all times. We are now witnessing the early incarnation of this with our increasingly advanced cellular phones.[7] They have become powerful tools, which not only link us to other people, but also our environments. They have become portable computers that allow us to instantly access information about our surroundings. Although the impact that smart phones have had on the world is impressive, this has been only a glimpse of what is to come. As people become extremely connected by both real and virtual worlds, there will be dramatic residual effects on the individual personality. The relationship between individuals and society as a whole could become far more intimate than it is at present. "Hive mind" is a term used to describe the convergence of the whole of humanity into one "interconnected virtual brain." Going beyond mere communication, the end goal of the hive mind is actually to enhance our intelligence. By merging disparate minds, human intelligence could be united in a way that was never before possible. Connected minds would presumably share ideas and information at dramatically higher speeds than was ever possible before. Once separate minds could become a globally interconnected super intelligence. This would of course enhance human performance

---

[6] *NBIC*, page 6

[7] The report details a device deemed "the Communicator." Current cellular phones have some of the same features.

by revolutionizing our intelligence and knowledge. Although the hive mind technically exists in a primitive form today as Internet, such things as brain-machine interface, implantable chips, and other technologies could change things significantly by directly linking our minds together.

Being the C in NBIC, Cognitive science is indeed a very important aspect of convergence. Our brains drive our very existence within this world. The hive mind is just one of many proposed ways of enhancing human cognitive performance. Brain machine interfaces (BMIs) and actual computer microchips that would be implanted into the brain are also outlined at length in the NBIC report. Implantable chips could be used to store memories, speed up our thought process, and even increase our mind/body coordination. These chips could also act as the connection between the individual and the greater collective hive mind. Imagine Internet wired directly to your brain.

Just as the Human Genome Project laid out the basic genetic structure of the human body, so too could the "Human Cognome Project" reveal the intricate structure and functions of the human brain. This hypothetical project was proposed during the conference as another possible avenue to take to gain greater understanding of the mind. The intense study of humanity from the microscopic level of DNA all the way up to the macroscopic level of society as a whole is crucial to establishing total efficiency. This overarching perspective is described as the "continuum of bioinformatics." Total understanding makes total control possible, and make no mistake, that is the reason we are currently on the technological fast track: to establish complete control over all life on earth.

"Predicting social behavior" can be done scientifically. Although this is nothing new, the process could be perfected technologically. On page 140 it is stated,

> The multiple drivers of human behavior have long been known. What have been missing are the theoretical paradigm and associated tools to integrate what we know about these drivers into an overarching understanding of human activity.

In other words, the knowledge of our instinctual drives and urges has been understood for quite some time. With the advent of computers the entire game of decoding human behavior has changed.

"Socio-tech" is described as "the predictive science of societal behavior." It is based on data derived from the social sciences. The ability to collect massive amounts of data is possible with the creation of huge IT databases. Artificial intelligence could be created for the specific purpose of interpreting this vast amount of data and converting it into accurate predictions about any given person's future actions. This is the goal of socio-tech; to predict everything. Prediction is taken to extreme limits as one NBIC vision of the future has to do with the prediction of crime. The ultimate goal is to transform law enforcement by setting it toward the end of crime prevention. By monitoring people intensely, and understanding their psychology, artificial intelligence would be able to predict when specific persons are about to commit a crime *before* they actually do it. The future would be seen in advance and instead of letting a crime happen, criminals could be stopped before they had the chance to commit their crime.[8] Forget about the old idea of guilty until proven innocent. This method would suppose guilt before you even do anything! The NBIC report explains this as a wonderful way to prevent "terrorism." It all seems too bizarre to believe, but consider these words,

> Most immediately socio-tech can help us win the war on terrorism…In the longer term, as a predictive science, socio-tech can help us identify possible drivers for a wide range of socially disruptive events and allow us to put mitigating or preventative strategies in place before the fact.
>
> Socio-tech…will raise our ability to predict behaviors. It will allow us to interdict undesirable behaviors before they cause significant harm to others and to support and encourage behaviors leading to greater social goods.[9]

National Security is a major concern when it comes to this kind of development. The effect that this will have on our freedoms (what's left of them) will be enormous. This report even goes as far as to say that "certain ideas may have the force of a social virus," and so the need to understand what makes people believe "weird things" becomes necessary. This is incredibly disturbing.

Representatives from NASA and DARPA were on hand to discuss some of their many NBIC projects. Nanotech has been a priority research program for the DoD since the early 1980s and NASA has also developed a great deal of nano-enhanced materials. Based upon the 2001 level of development, several visions of the future were made in regard to military application of nanotechnology. The ability to create remarkably strong and light materials could eventually lead to the creation of "polymorphic" substances, which could structurally change at an atomic level on command. This would amount to true-

---

[8] This was the main theme within a major Hollywood film by Steven Spielberg.

[9] *NBIC*, page 140, and page 142. If you read one section of NBIC, read this one.

life shape shifting. The use of such technology has nearly unlimited military applications. Of course, all of these enhancement technologies have military applications, and that is why so many of them have been funded and developed through the DoD.

Unmanned Aerial Vehicles (UAVs) could literally have a mind of their own. AI could guide these military machines. In the past, fighter jets have been designed with the human pilot in mind. Once humans are physically removed from the cockpit, performance can be dramatically increased. It is also possible that an "enhanced" human could pilot such vehicles from far away, just as if they were controlling a video game. The most likely situation is that AI would control these vehicles with unflinching technical precision. Human error would be deleted. The mind and body of military aircraft could be intimately connected to its specific tactical missions. The shape of these craft could take on that of a living bird. Birds were explained as one of the inspirations for new vehicle designs. These new craft could be alive in a very real sense. Their autonomy would be absolutely staggering. UAVs could also maintain themselves.

Not only were military birds of prey discussed, there were also many visions for improved ground forces. "The high performance warfighter" could be created using a wide variety of NBIC tech, from pharmaceuticals to cybernetics. The ability to create a soldier that never sleeps and consistently performs at his peak, both mentally and physically, was seen as a realistic goal. Various methods of affecting the brain could be used to increase performance. Drugs, electromagnetic stimulation, and of course brain chips, could keep soldiers fighting. Brain-machine interfaces would integrate the soldier with his equipment to create the ultimate fighting machine. New kinds of body armor that not only protect against damage but also enhance the muscle power of a soldier could be developed as well. The high performance warfighter is truly an amalgamation of man and machine.[10] Again, the possibilities are endless.

Genetic engineering and organization was envisioned at a global level in this report. "Ideally all life on Earth would be catalogued," the report stated. This is the ideal situation for an organizational body obsessed with managing the health of citizens at a global level. It is not, however, the ideal situation for an individual who values his own privacy. That is why people must be trained to accept these

---

[10] The ideal soldier has always been automated. Stripping individual will, and the establishment of strict regimen has always been the first step in programming humans for warfare.

technologies. It is hoped that we will continue to believe that technological development occurs for the mutual benefit of everyone. After all, medical treatment would be dramatically changed by NBIC. "Nano-bio-sensors" could probe any living organism to provide complete and total information on that specific creature's genetics and state of health. No doubt, diagnosis of illness would be far easier, but privacy would be eradicated. This would admittedly "enable big-brotherism," which is not necessarily a bad thing according to this report. Current laws are a hindrance to establishing such a progressive world of "biosurveillance."

Newt Gingrich attended the NBIC workshop. His stirring speech got the gathering fired up. He predicted a future which would be profoundly affected by the convergence of technology. According to Gingrich, the coming wave of dramatic technological developments and cultural changes would be too much for most people to handle. Society would find itself in a state of flux as rapid paradigm shifts would require that government be able to quickly respond to various threats in an intelligent fashion. It would have to organize chaos in a time of rapid change. The "age of transitions" as he called it, would require bold action, and more importantly, advanced preparation. In the meantime he suggested that massive funds be allocated to technological development insuring the U.S. stay ahead of the curve. He warned that any country that chooses not to adopt these coming changes would become "weaker, poorer, and more vulnerable than their wiser, more change oriented neighbors." Government is pushing the tidal wave, which is Singularity, and it is a truly global force. Is it any surprise that member of the NSF, and editor of this very report Mihail C. Roco said that, "globalization is a key goal." It always has been. Technology is a means by which government may obtain wider command and control power.

With the global aspect of this project in mind let's look across the Atlantic Ocean to see how Europeans have planned for emerging technologies. The United Kingdom Ministry of Defence (MOD), in preparation for a chaotic time of massive change, has written up a similar document to NBIC. The *DCDC Global Strategic Trends Program 2007-2036* outlines its predictions for the future and gives some suggestions as to what the British government should do to plan for these coming events. The future envisioned is incredibly bleak as it foresees crisis after crisis contributing to global instability and intense competition for survival.

Amongst the threats of climate change, food and water shortages, pandemics, high crime, urban warfare, terrorism, decreasing fertility, and "unexpected shocks" there will also come a dramatic wave of technological innovation. The report boldly predicts that every aspect of human life *will* change. In the beginning they make a distinction between the words *will*, *likely*, and *may*. These are the varying degrees to which they gauge the accuracy of their predictions. It is known for certain that every aspect of human life will change. This is an incredible prediction. How will these changes come about?

The report goes on to explain that a blend of total chaos and technological convergence will shock people into a state of utter confusion. This will cause them to become reactionary as well as question their previous belief systems. When put into a survival situation, people will come to accept new technologies by default. Scientific materialism will become increasingly more influential as fierce competition for survival will transform new technologies into survival tools. "Flashmobs" will routinely break out in crowded urban areas, and in an effort to deal with this threat, the military, police, and private security firms will join into a singularity of their own. "Security forces" will constantly patrol the streets, many of which will be made up of foreign military troops and private mercenaries.

As chaos ensues, the trend of increasing interconnectivity between people and their communications devices will continue. Individuals who create their own "themed websites" and contribute to the "blogosphere" will threaten the mainstream media. There will be an intense infowar over public opinion as truth will be contingent on what is "believed" rather than what is actually happening. Control over communication will undoubtedly be a top priority as people are literally wired to news feeds. The creation of implantable chips relates to this and is described in chilling detail.

> By 2035, an implantable information chip could be developed and wired directly to the user's brain. Information and entertainment choices would be accessible through cognition and might include synthetic sensory perception beamed direct to the user's senses. [11]

A virtual reality could be substituted for true reality. These implantable chips could enable complete tyranny. Controlling the population and dealing with radicalized factions of "terrorists" would be much easier, or perhaps even unnecessary, if mind control devices were developed and successfully implemented on a mass scale. It could certainly lessen the demand for the many paramilitary security forces described here. Without a doubt these brain-chips will have many ramifications for individual privacy.

---

[11] *DCDC*, page 82

It is likely that the majority of the global population will find it difficult to turn the outside world off. ICT is likely to be so pervasive that people are permanently connected to a network or two-way data stream with inherent challenges to civil liberties: being disconnected could be considered suspicious,[12]

Much concern is raised over "suspicious" individuals and radicalized factions of "terrorists" who reject the mainline explanations of world events. More than this, it is expected that average citizens who previously held no strong political convictions will become revolutionary and act out violently. A total surveillance society is again given as the solution to solve the problems of managing a chaotic world.[13] AI systems will be developed to effectively manage governmental and commercial enterprises. As machines become more powerful, humans will increasingly enhance themselves with cybernetics in an effort to stay ahead of the curve, but one has to wonder if that would even be possible.

Strat Trends explains that the world will be globalized, and broken up into a very disturbing system of social stratification. There will be those who have access to high technology (the owners), and those who do not (everyone else). There will be:

> a small super-rich elite and a substantial underclass of slum and subsistence dwellers.[14]

Beyond this, there will be social warfare occurring between generations as old cultures are steadily worn away. There will be animosity between parents and children and the clashing of ideologies will lead to the breakdown of long held traditions and values in an increasingly uncertain world. It appears as though a great divide and conquer scheme is being set up in an effort to manage the masses as technology is put on the developmental fast track. The question we all need to ask ourselves is: is this terrible future actually desired by the current "super-rich elite?" If so, are they making preparations now in an effort to purposely reshape the world according to their interests?

In many ways, the entire push toward Singularity seems to be an effort to scientifically predict everything and to establish total efficiency. "Enhancing human performance" is not so much an effort

---

[12] *DCDC*, page 58

[13] Most of these predictions have already occurred to an extent. Internet itself has already been deemed dangerous, and plans to restructure it have been ongoing.

[14] *DCDC*, page 78

to help people as it is a way of streamlining society as a whole. These reports clearly state the desire to understand complex systems. What are the most important complex systems to understand? Society is a complex system made up of humans. Humans themselves are complex systems with their own complex systems of biology (nervous, circulatory, reproductive, etc). We see the mystic macrocosm to microcosm analogy in this "continuum of bioinformatics." Through advanced bioinformatics research the human body and its many systems could be understood, but more importantly, they could be altered. Genetic engineering would make more efficient (and predictable) humans, which could be utilized to bring about a more efficient society.

The MOD report describes "combinatronics" as the effort to decode complex systems. To do this requires the ability to recognize patterns. The ability to collect massive amounts of information, intelligently filter recognizable patterns from the info, and then to understand the significance of those patterns is the idea. This will enable one to make accurate predictions based on a history of patterns. Combinatronics is the technical way to achieve predictability. With advanced combinatronics one could obtain greater levels of control over all forms of life on earth. The "combine" aspect of the word also alludes to the convergence of interests that is bringing about this vision. A combination/convergence of government, corporations, NGOs, academia, and powerful individuals is the best way to reach such high goals. The convergence of different sciences and technologies makes understanding complex systems easier.

Plain and simple: this is about understanding everything. There is no big mystery here. The only thing causing confusion is that most people have not been presented with the facts. They haven't taken the time to look at this beyond the scope of repetitive news blurbs.

So how does such a massive project actually work? How could anyone be able to understand everything? A good first step is holding conferences and writing up plans, but to truly pull off such an ambitious vision requires something more. It requires the creation of superintelligence. Herein lies an incentive to create artificial intelligence. A powerful artificial general intelligence would not only be able to recognize and interpret massive amounts of patterns, it would also be able to make informed plans of action based on its profound level of understanding. This sort of intelligence could make sense of the human hive mind. It would understand everyone's identity as an individual and also as a small contributor to society as a

It's Official

whole. It would be able to interpret the "continuum of bioinformatics." From our limited human perspective such a superintelligence could be considered omnipotent, a godlike power on earth. This just might be the idea.

Artificial intelligence is the key to opening the door to Singularity. That is why so many important entities are hell bent on making AGI a reality. The Singularity Institute for Artificial Intelligence (SIAI) is a major group dedicated to this task. It should be noted that the SIAI is located at NASA's Ames Research Center (ARC). Its top goals include the testing of AI at a "baby" level from 2007-2011, and afterwards moving on to further development. Beyond AI specific research, the institute is one of the highest profile groups advocating the technological Singularity.

Technological progress is an awesome force. The reality of convergence is not well understood by the common man, and yet it becomes increasingly important to his everyday survival. The development of AGI has already started with companies such as Novamente, which works with and is financed in part by the SIAI and Google. The SIAI itself is also funded in part by Google, a company that has expressed open interest in AGI many times. As for Novamente, its owner, Ben Goertzel, is the director of research at the SIAI and he also sits of the board of directors of the World Transhumanist Association. Novamente has impressive clients including the Center for Disease Control (CDC), Northrop-Grumman, NIH Clinical Center, the Global Health Exchange, and more. Both the SIAI and Google have contributed funding to Novamente and have facilities at NASA's ARC (Ames Research Center: it's a bit like Noah's Ark, but higher tech). This is a team effort of epic proportions.

In addition to Novamente and the SIAI, many projects are being conducted by DARPA that have to do with AI and its many related fields of research. The amount of programs is staggering when you actually look at them. It is amazing to see how many separate initiatives are on the go at the same time, many of which would sound the same to the casual observer. Let us not forget the many other alphabet agencies which are cloaked under the veil of national security. Google has recently come out openly expressing that it works in conjunction with the NSA. There is no way of telling how long this partnership has been going on. Without question, the desire to create AGI exists. There is a very real public/private partnership that drives overall development and much of it is public knowledge. Hundreds of billions of tax dollars are doled out annually to the NSF, the DoD, and all other grant giving agencies of our government. If you want to develop any advanced technology, then you have to procure a grant from one of these resources.

Converging technologies are not emerging from out of nowhere. They are the product of trillion dollar budgets. The amount of money put into federal departments such as the NSF, DOC, NIH, and DoD is absolutely staggering. Advanced technologies will enable their owners to acquire real wealth. This is why an endless supply of fiat currency is produced to fund development. The legislative branch of our government creates initiatives and alongside the executive branch approves the massive budgets of every large department annually. The money used to fund these programs comes directly from the Federal Reserve Bank, which sits above and beyond our government. It is a private bank that issues as much currency as is requested, but it does so at a cost. Interest is charged on every dollar issued by the Fed. *All* dollars are created with interest attached, and so we see that the debt accumulated by the mere creation of money is impossible to repay. At least, it is impossible to pay back in dollars. There can never be enough to make up the interest. What will be the ultimate cost paid by the American taxpayer in order that the Federal Reserve's interest is appeased? This is the true issue that transcends all politics. The interest that is impossible to pay back is merely a guarantee that the Fed's interests are secured.

The deficit spending of our government is fulfilling multiple purposes. One is to see that real tangible assets are produced as fast as possible. Converging technologies represent true wealth and power. Whatever debt is accrued on the way to technological dominance is the burden of the borrower. In case you missed the point, let it be restated: you are the borrower! This is where we see the other reprehensible purpose of deficit spending: to get the taxpayer dependent and hopelessly tied to a debt they can never repay. The convergence of these trends brings about a truly disturbing angle never discussed by advocates of the Singularity. Our government is owned through fiat by private banks. Anything produced with the dollars issued by the Federal Reserve Bank may be taken as collateral on the debt accrued in the process. More than this, we have to understand that the power of converging technologies absolutely will be used to maintain this entire system. Even if the outer appearance of our government changes radically, we must understand that the debt owed to the Fed never disappears. It will be paid one way or the other. Someone has set up a massive pyramid scheme in order that they may become owner of the world and everything in it. If you

It's Official   53

haven't done so already, you must research the Federal Reserve system. A wonderful film that can get you up to speed on the subject is Aaron Russo's *America: Freedom to Fascism*.

To view the various institutions that are driving scientific advancement as separate entities would be a mistake. It is clear that convergence is a team effort. The very term *singularity* is a nod and a wink to those working within the confines of this massive apparatus, which could be (and is) described as a living thing in its own right by many experts. The creation of a new and improved global system requires teamwork. We know that early versions of AI are being put to work on bioinformatics, to better understand our biology, but what happens next? Will a powerful AGI ever be created? Just how far along is AI research already? If a super intelligent AGI existed, would it purposely remain in the shadows? Don't write off that last question, because it is valid. We live in a world that operates on a need-to-know basis. Most of us are on the bottom rung of the ladder, and consequently we know nothing.

There is something undeniably dark going on here. The average person can pick up on this fact and they visibly express their feelings of anxiety and unease when any hardcore transhuman topics come up. Most people don't like the idea of implantable chips, genetic engineering, or superintelligent AGI, at least not yet. But as the white papers say, attitudes change.

To say that it is merely irrational fear that causes people to question the desirability of the Singularity is not fair. Singularity presents a true danger to the human race, and the most renowned Singularity advocates freely admit to this fact. AGI already has its own cult following. This cult mainly consists of researchers who are searching for a way to bring their god to life. They are operating on the belief that the possible benefits of creating AGI outweigh the dangers. Yes, humanity could be, and likely will be eradicated, but something better will take its place. The attainment of superintelligence is well worth the cost. This is something that is consistently stated by the AGI cult.

Understand that it would be impossible to implement the transhuman vision and remain completely human. A posthuman would be so different from us that it could not be considered the same species. The end of life as we know it is the only thing that we could be certain of upon entering a posthuman world. Now if this isn't a dark concept, then I don't know what is. The transhumanists have a huge job to do. First and foremost they are tasked with convincing the world that this is all a good idea. Doing so requires a masterfully crafted detour of the human survival instinct itself. Is this the truth behind transhumanism? Have our subconscious feelings of dread been correct? Is the human species endangered, and if so, then who is to blame? These are hard questions to answer, but they have to be asked. We cannot glaze over concerns for the wellbeing of humanity in a mad rush toward a technologically driven chaotic future. If it is possible, then we must stop and think about where we are and what we are doing. Now is the time to ask all of the hard questions that have consistently been ignored.

-Chapter Three-
Public Relations

> Propaganda assists in marketing new inventions. Propaganda, by repeatedly interpreting new scientific ideas and inventions to the public, has made the public more receptive. Propaganda is accustoming the public to change and progress
>
> -Edward L. Bernays, Propaganda, 1928

How do you go about enhancing people, turning them into posthumans, if no one is interested in the first place? Most people would not want transhuman upgrades outright; and that is why so many public voices are hard at work trying to convince us that this is all a good idea. Public acceptance must be manufactured[15] before any actual technologies are distributed. To do this effectively requires the intelligent creation of propaganda. Wars bring about major societal change and, truth be told, there is an ongoing war to win over the hearts and minds of the people. This war is waged in plain sight and most never even notice it. Folks even go so far as to steadfastly deny that such a war even exists due to a lack of evidence. All one needs to do is open his mind to see that he is swimming in a sea of evidence. Propaganda has forever been a force at work within civilization. The larger the society, the more important it has been to create psychological cohesion. Standardization of purpose is established by directing minds unconsciously. The hallmark of true propaganda is that it goes completely unnoticed due to people's ignorance of what it actually is. It must be taken for granted. In his book *Propaganda: The Formation of Men's Attitudes,* Jacques Ellul said that,

> only with the propagandee's unconscious complicity can propaganda fulfill its function.[16]

Unfortunately, Ellul was correct.

---

[15] In addition to *Propaganda*, refer to *Engineering Consent* by Bernays.

[16] Ellul, *Propaganda*, page 160

In the NBIC report's section entitled *Memetics: A Potential New Science*, Gary W. Strong and William Sims Bainbridge of the NSF explain the importance of social engineering,

> In the information society of the twenty-first century, the most valuable resource will not be iron or oil but culture.[17]

Culture, when properly understood, could be shaped with direct purpose,

> Debates range over how heavily government should be involved in supporting culture through agencies like National Endowment for the Arts or National Endowment for the Humanities. But here we are discussing something very different from grants to support the work of artists and humanists. Rather, we refer to fundamental scientific research on the dynamics of culture that will be of benefit to culture-creating and communication industries, and to national security through relations with other countries and through an improved ability to deal successfully with a wide range of nongovernmental organizations and movements... If manufacturing creates the hardware of modern economies, the culture industries create the software.[18]

Culture creation is correctly described here as an industry. What we blindly accept as culture has been professionally crafted and given to us in order to fulfill our basic human need for social identity. Culture creates identity, and industry creates culture. This means that our very identity is the product of industry. To a great extent, our minds have been standardized like products on an assembly line. Any degree of escape from this corresponds directly to the strength of our individual will. Be sure to note that the above quote explains that the culture creation industry is intimately connected with national security, NGOs, and various other "movements." Now, what movement is concerned with creating a culture that is tolerant of a post-human future? The transhumanist movement obviously fulfills this role. Could it be that transhumanists work directly with the governmental and private organizations in attendance at the NBIC conference, and that all of these parties work together in a directed and deliberate effort to gain public support for emerging technologies?[19] The transhuman enhancement meme is being put out

---

[17] *NBIC*, page 279

[18] *NBIC*, page 282

[19] William Sims Bainbridge not only works for the NSF, but is the editor of the NBIC report itself. He also happens to be a Senior Fellow of the IEET, which basically *is* the World Transhumanist Association. The IEET's co-founders have both been Executive Directors of the WTA. One of them, Nick Bostrom, *openly* admits to his consulting work with the CIA.

Public Relations

by the Singularity, which is government, industry, NGO, and academia. This is an all-encompassing agenda. The WTA itself doesn't require massive funds to be effective, because it is only one small portion of a larger whole. It could easily disappear and its ideas would carry on.

Decoding complex systems becomes easier all the time due to advances in socio-tech but public relations has been a science from its very inception. Its best practitioners have understood this well. They have known that through studying human psychology one may learn how to exploit and direct human behavior. It is possible to coerce people into doing anything if you have the proper technique. Edward Bernays, the self proclaimed father of public relations, was a master of this psychological science of coercion. It should come as no surprise that he was the double nephew of Sigmund Freud. It seems the family business was psychology.[20] Bernays was a dark genius, a master of his craft. It's quite fitting that in our twisted world this man is held in extremely high regard by public relations professionals. His book *Propaganda* is standard curriculum within academia. Public relations students are introduced to a world that they often don't fully reflect upon due to their own professional diligence. Bernay's ideas are scientific and, as such, moral judgments of them are cast aside. The ethics involved in scientific research are often unclear but what has become obvious is that mass mind control is now a well-established and legitimate business. It goes by the name Public Relations. The following is from *Propaganda*,

> If we understand the mechanisms and motives of the group mind, it is now possible to control and regiment the masses according to our will without their knowing it.[21]

The "we" that Bernays refers to here is of course PR professionals and their clients. Big money is now pumped into this industry, the biggest money.

It is important to understand that Bernays identified himself as one of the "intelligent few" who have the privilege of directing society. This is evident from reading *Propaganda*.[22] There can be no doubt that Bernays knew that an intellectual elite purposely guides society according to its own blueprint, and that he counted himself as

---

[20] In fact Bernays learned much about psychology from his uncle's writings.

[21] Bernays, *Propaganda*, page 47

[22] A good read, and first hand account of this point is also found in Chapter 1 of Stuart Ewen's *PR! A Social History of Spin*.

one of them. He was a true insider within the world of history creation, a key player in the world unification/world democracy movement that spanned the entire twentieth century, and has led up to today. We have not arrived to where we are today by happenstance. A small group of powerful individuals formulate agendas so that their will may be done. Propaganda is simply the means by which these agendas are promoted. Such a method may take a long time but it works none-the-less. The public is eventually convinced to willingly, albeit unconsciously, carry out the will of the intelligent few. Before denying this unpleasant reality, you must first read *Propaganda*, by Bernays himself. He states the truth bluntly, and the only way around his numerous confessions is through willful denial,

> The conscious and intelligent manipulation of the organized habits and opinions of the masses is an important element in democratic society... those who manipulate this unseen mechanism of society constitute an invisible government which is the true ruling power of our country... it remains a fact that in almost every act of our daily lives, whether in the sphere of politics or business, in our social conduct or our ethical thinking, we are dominated by the relatively small number of persons... who understand the mental processes and social patterns which control the public mind, who harness old social forces and contrive new ways to bind and guide the world.[23]

Uncomfortable truths must now be faced courageously before this process becomes technologically perfected. We must know that propaganda, PR, socio-tech, and memetics are indispensable tools used with military precision to mold public opinion and establish social order. Socio-tech and memetics are new terms used to describe an increased scientific understanding of mass psychology. The mass mind has existed throughout history. Wherever civilization was known, it too was there. It is this hive mind that has been studied by technocrats down through the ages. By understanding the motivations of the group, one can deduce techniques to influence its behavior. Entire societies have been directed to a great extent, and with chilling accuracy through primitive techniques. Just think of the possibilities that are now arising due to converging technologies.

As the NBIC report states in its socio-tech section, advanced technology can enable large-scale surveillance and prediction,

> The time is ripe to begin such integration - to use the tremendous computing power we now have to integrate data across these fields to create new models and hence new understanding of the behavior of

---

[23] Bernays, *Propaganda*, pages 9-10

individuals. The ultimate goal is acquiring the ability to predict the behavior of an individual and, by extension, of groups.[24]

The report goes on to describe specific advances in neuroscience that are leading to a more complete understanding of the human brain.

Our basic human motivations, desires, and behaviors can be broken down into patterns. By thoroughly understanding these patterns new ideas can be introduced in such a way that isn't disruptive to them. Again, this is all about pattern recognition, something that AI developers love to go on and on about in their publications. First you recognize a pattern and then you may expand upon it. This act of building is equivalent to engineering. Social engineering involves the creation of new ideas. These ideas must take hold within the mass mind in order to gain widespread adoption and this is where propaganda comes into play. All propaganda, public relations, and marketing must be congruent; it must fit within the larger social patterns that give a particular civilization its identity.

A perfect understanding of mass psychology has always been elusive to the extent that its practitioners could never account for every single anomaly that would happen to arise from within the population. From out of nowhere something unpredictable could occur. Spontaneity can never be fully accounted for. The dynamic human spirit is an obstacle to perfect legal order. To bypass this spirit (that of freewill itself) is a high esoteric achievement. This has always been the holy grail of the tyrant. Complete and total world domination has never quite worked out, and for good reason: The mass mind, and even the mind of the individual may be well understood, but how does one keep order in a world where the slightest impulse can lead to untold cascading effects? This wave is the nature of chaos, something despised by control-freaks.

Yet another unpredictable wave is rising today and every effort is being put into taming it. The wave of technological advancement has but one sure end: total change. Although the final results of this change cannot be completely known until they occur, serious effort is being put into directing this force of *nature*.[25] There exists a plan to bring about a cybernetic merge of man and machine. This is desired for many specific reasons and in fact these reasons are not fully understood by the majority of people who are the actual focus of such

---

[24] *NBIC*, page 141

[25] The term *nature* is used loosely here. Within mystic philosophy it is believed that technology is one of the highest products of Nature.

change. We are given endless praise of technological progress and optimistic visions of a transhuman future, but what about that nagging voice of doubt that continues to beckon in the back of your mind?

The transhumanists speak of a world liberated by converging technology. They tell us that by taking implantable chips and other cybernetic add-ons we will be upgrading the human condition itself. Criticizing such technologies is portrayed as an irrational and reactionary response. A deliberate attack is made on all those people foolish enough to believe that technology is now at a dangerous level of advancement. This keeps the multitude of timid doubters at bay. Transhumanists cunningly use persuasive language such as "upgrade," "enhance," and "transcend" to make their product sound beneficial. A good example of this public relations tactic can be seen in the WTA's official name change of 2008. This is when the World Transhumanist Association officially changed its name to Humanity Plus, or H+ for short. It even came up with a new slogan, "better than well."

In the *About Us* section of its website, the first subsection reads, "What is the Mission of Humanity+?" The answer given is, "We support the development of and access to new technologies that enable everyone to enjoy better minds, better bodies and better lives. In other words, we want people to be better than well." It sounds so wonderful, doesn't it?

The label *transhumanism* isn't very friendly or approachable. Beyond being hard to pronounce, it is actually quite disturbing. It presents a fundamental challenge to the human race by offering an alternative life form and this is threatening. The move away from humanity itself is implied in the structure of the word transhuman. It is not easy for most people to relate to this but they can easily respond to the rhetoric self-improvement. Think of how many self-help books exist. Each one of these books promises the "secret" to happiness. Under the banner of Humanity Plus, the transhuman campaign does just this. It is delivered as a means of improvement and not abandonment of our unique human identity. It implies that we are adding onto ourselves, as opposed to mutating into something entirely different. This is a much more user-friendly label for an idea that is meant to go completely mainstream.

Through PR techniques we are expected to believe that the transhuman idea is truly about improving the human race and nothing more. We aren't supposed to realize that efficiency is the goal of those who are developing converging technologies, and that the individual human mind is a hindrance to these ends. We aren't supposed to be conscious of the fact that we have been bombarded by a campaign of psychological warfare our entire lives. Bernays constantly referred to propaganda as a "weapon," and with good reason.

It is no exaggeration to say that psychological warfare has been waged on us. In the year 2008 the Associated Press ran an investigation determining that the U.S. military spent $4.7 billion in the course of one year on "winning hearts and minds at home."[26] The Pentagon pays PR services to distribute "news" stories across the country. It employs 27,000 people for military recruitment and PR purposes.

An information war is being waged against us and we don't even know it. The Pentagon officially describes its psychological operations (PSY-OPS) as targeting foreign audiences. Our military sets up media installations in occupied countries and trains locals to use them on its fellow citizens. This is how, through coercion, foreign nationals are made to see things the American Way. PSY-OPS are used against enemies to achieve victory. The military as a whole is used to carry out missions. They destroy opposing forces which stand in the way of the overarching goals of "our" government. At home, and on far distant shores, these goals are ambiguously defined to the public at large. Quite simply this is done to get people on board with the mission. For reasons of national security the true goals of the military are never completely divulged to the masses of any country, foreign or domestic.

PYS-OPS mold the enemy's mind abroad. An apparently separate division officially referred to as "public affairs" works at home. Public affairs are the Pentagon's domestic psychological operations, and they are a necessary aspect of winning its wars. The same AP investigation found that $547 million were devoted to this cause in 2008. Beyond this, it detailed the fact that former Secretary of Defense, Donald Rumsfeld, established an Office of Strategic Influence that officially brought together both the PSY-OPS and public affairs divisions. In this we can clearly see how all $4.7 billion of this particular budget (involving both PSY-OPS, and public affairs) are effectively used for one global propaganda campaign. Rumsfeld explained that the American public should realize that with major media outlets operating at a global scale, it is only natural that foreign and domestic PSY-OPS would merge. He said this in 2003

---

[26] *Pentagon Spending Billions on PR to Sway World Opinion,* Associated Press, Thurs. Feb. 05, 2009, foxnews.com

within a secret *Information Operations Roadmap* that detailed how the two previously separate divisions could work together.

Presidential administrations are easy targets of public condemnation. It is all too easy for us to point our collective finger at a single figure. Characters like Donald Rumsfeld and George W. Bush, while being guilty of certain atrocities, remain scapegoats. Most people don't care to see the cohesion between their particular unconstitutional acts and the many new ones that are forthcoming from each new administration. Keeping the ball rolling has been standard operating procedure with our government for quite some time now. Each new presidential administration builds upon the foundation laid by its predecessors. If we are to become truly aware of what has happened to our country, and our world, then we must understand this. We have to see through the petty PR news stories that are distributed en masse to give us the illusion that things are changing for the better. Things are changing all right; they have been for a long time, but how, why, and for what? Start asking real questions. Start questioning the very definitions given for powerful words such as democracy.

The transhumanist marketing campaign, which is designed to sell the idea of a technoprogressive democracy, falls apart when we understand how democracy actually works. The very word *democracy* has been built up by a century-long propaganda campaign which has been led by psychological technicians such as Edward Bernays. Democracy is a meme which has been constantly repeated and generically defined. Its power of ambiguity cannot be underestimated for it is exactly this that allows full spectrum control across entire populations. Everyone can jump on board with ideas of mass appeal. Words such as democracy become polymorphic, for they have the ability to shape-shift in order to cater to a particular person's values. The average man[27] holds his own unique definition of democracy that conflicts with that of the person standing next to him, and yet the majority agrees that democracy is a good thing. The bulk of society remains content, always believing that its particular interests are being catered to. The true government remains in the shadows as it keeps the masses happy with the illusion of freedom.

From a purely technical point of view, manipulating the mass mind is a completely necessary act of organization. Large societies need to be unified. The population must tolerate the system itself. The complex system that is society requires a never-ending public

---

[27] Ellul accurately refers to him as the mass man.

relations campaign to keep the populace content and obedient. If it weren't for propaganda, then people would constantly be halting progress with their nagging questions and doubts. The true progressives in this world just can't have this.

Crucially important to remember is that the WTA is the *World Transhumanist Association*. In addition to memes involving implantable chips, AI, life extension, virtual reality, and nanotech, the WTA also strongly advocates the "global governance" meme. They often make the claim that global governance is necessary in a world of increased complexity and interconnectivity. Transhumanists say that within a world democracy, technologies would not be hoarded by an international elite, but rather they would be generously given out to everyone to transform the world into a progressive wonderland. Citizens would be cyborgs and government would be global.

James Hughes, former executive director of the WTA and author of the book *Citizen Cyborg* has stated,

> We build global citizenship when we focus on the need for global governance to address global threats, and provide global affluence.[28]

Was the word *global* used enough times in that sentence for you to remember it? Hughes is a major proponent of the technoprogressive view of politics which has become the predominating stance of the transhumanist movement in general. At one time there was a strong left/right political dialectic among transhumanists, with technoprogressives on the left, and extropians[29] on the right, but now the right has seemingly fallen back as the majority of prominent transhumanists take the technoprogressive point of view. This very well could be a deliberate effort to subdue public fear of the movement by making it seem more beneficial to the common person. Technoprogressives speak of a world in which life extension, cognitive enhancement, and nanotech will be distributed to everyone in order to benefit global society and "eliminate scarcity." They say that technology must be "equitably distributed across all strata of society." They believe that government has an obligation to serve its citizens in a fair and democratic manner. This sort of consideration is comforting to anyone who trusts that these technologies will in fact be beneficial as advertised, however a major problem remains. That problem is the potential malevolent use of emerging technologies. What would happen if these things fell into

---

[28] *Choosing Our Imaginary Communities and Identities*, May 18, 2009

[29] Extropians are quick to claim credit for starting transhumanism in the 1980s. The "fact" that they started the movement is debatable to say the least.

the wrong hands? An even better question to ask is: what happens if these technologies are already in the wrong hands, and are in fact being developed for malevolent purposes? Convenient answers to these questions always seem to be provided by the transhumanist movement itself. Is this done in order to bolster an unhealthy trust in people toward their government? Are the real questions being brushed aside purposely?

"Making the world safe for democracy." This is simply a slogan, a catch phrase, a marketing technique, but what it has wrought is truly awful. Real violence has been enabled by a clever use of language. Sweeping generalizations are part of a military strategy to win hearts and minds. The American people have had to be convinced that wars are being fought for a noble cause, or at the very least distracted enough that they don't demand real answers. We falsely believe that the US military is our military, but in reality it is already the world military. The reason that it is "bringing democracy" to the world is to standardize the world's governmental/legal system. Global governance is being established, and we are going to like it. We are being hit hard by military-grade psychological warfare. This is being done in an attempt to reorganize the globe, and upgrade it, to reorganize everything into one highly efficient machine; a one world governmental structure.

There is no way that a global system of government could ever come to fruition without the consent of the world's population. Brute force only goes so far. What people really want is peace and freedom, and this is what they will get. They will be offered the world, more specifically, a better world, a world beyond scarcity, where every human need is catered to, where health care is revolutionized and freely given to all, a world where the aging process is actually reversed and life extension is given to everyone, a world where all hard labor is performed by machines allowing humans the freedom to spend their days as they wish. This is the technoprogressive global democracy of tomorrow, a world society that will do away with competition and reign in peace. It sounds wonderful because it has to; otherwise no one would want it. This vision of a world united is the largest propaganda campaign at work in the world today. This campaign is far beyond the realm of H+; it is everywhere you look. Media is being used to advance this message. When we understand the scale of communications media, it becomes clear that we are now witnessing the biggest psychological operation in the history of the world.

It is no accident that nearly every policy piece that you find regarding the creation of emerging technologies mentions the need for global governance. This "need" absolutely requires a massive propaganda campaign and this fact is often alluded to in the policy plans as well. We have been bombarded by a propaganda campaign promoting world unification our entire lives. In conjunction with this massive psychological operation we have seen very real geopolitical power moves made. Simply review the past century. If you remember, after each world war there has been an international organization set up specifically to carry out peaceful negotiations and establish new world standards through treaties. The League of Nations and the United Nations were two incarnations of one global project. Would you be surprised to learn that Mr. Bernays traveled with Woodrow Wilson to Versailles to play his part alongside the League of Nations? Never mind the fact that prior to this he had worked for the U.S. military's Committee on Public Information (CPI), which was responsible for producing pre-war propaganda. The CPI was tasked with convincing a wary American public that the war was necessary. This was actually the group responsible for producing the line "Making the world safe for democracy." By the time the war was over, Bernays was still quite young but he had learned a lot. He saw how powerful slogans could be used before and after wars so that the larger, more important and unseen war could continually be waged. He continued using the "weapon" of propaganda throughout the twentieth century to convince people of the virtues of supposed democracy. So we can see that big players on the world scene use both war and peace to fulfill one agenda. They are the masters of a dialectic method. As in Orwell's fictional *1984*, "WAR IS PEACE."

Everyone inherently understands the many dangers involved with creating "human enhancement" technologies. It would be foolish to develop them without any oversight or planning. Because there is a need to address the ethics of these technologies, organizations such as the WTA (H+) greet the public. This is a prime example of the democratic society that Bernays outlined. Instead of actually confronting issues as a free society and having our unique voices and opinions affect the direction of government, we instead believe that organizations, think tanks, and non-profits truly look out for our best interests. Instead of harshly criticizing the premise of human enhancement, we allow the experts to explain the dangers to us. This is because we have systematically been trained to revere expert opinion and ignore our own survival instinct.

In an effort to deal with the ethical issues involved with transhumanism, the WTA created a subgroup known as the Institute for Ethics and Emerging Technologies (IEET). With projects such as the "Securing the Future Program," "Rights of the Person Program," "Larger, Better Lives Program," "Envisioning the Future Program," and "The Cyborg Buddha Project" we can rest assured that these experts are hard at work making sure we all benefit from a technological Singularity. The fluffy titles of their *programs*[30] serve to convince us (and actually the transhumanists themselves) that the entire transhuman project is a benevolent and wonderful thing.

Yes, transhumanism is a PR campaign, but it must be understood that many transhumanists are in the dark themselves. They sincerely think that they are doing the right thing. H+ itself is primarily made up of volunteers. To really understand the movement, one must look at the many individuals within it. The interests and affiliations of many top-ranking transhumanists intertwine with those of the larger apparatus hard at work converging all technology. Even these individuals are marginally important in the grand scheme of things. Realize the true magnitude of the Singularity to understand that the significance of any individual playing a role in its very creation is microscopic. Consider these quotes from Bernays, again from *Propaganda*, in which he talks about the level of understanding that most apparent social leaders hold,

> It is well known that many of these leaders are themselves led, sometimes by persons whose names are known to few... Our invisible governors are, in many cases, unaware of the identity of their fellow members in the inner cabinet.[31]

Our entire world operates on a need-to-know basis. That is why we would all do well to humbly admit our own ignorance so that we may begin to gather real knowledge of our situation.

A technoprogressive world beyond scarcity sounds wonderful. The problem is that this vision feeds off our ignorance of how the world actually operates. We foolishly rely on blind faith. To merely hope that we will be taken care of by government has become standard practice in our society. Propaganda has instilled this ideal. We have become very good at demanding benefits, but extremely bad at understanding their cost. We have forgotten how to interpret the

---

[30] *Program* is a very important word. Computers run on programs, television programs effect our opinions. To "get with the program" is to follow specific rules.

[31] Bernays, *Propaganda*, pages 33 and page 9

Public Relations

universal poison label: a skull and crossbones. Today we think such symbols look edgy and cool.

The mass media is now everywhere. In our day it is hard to escape, if even for a short while. Television, radio, internet, billboards, movies, newspapers, and magazines are all examples of media that permeate our lives. Reflect on the amount of hours a day you spend in the presence of these things. Escaping the influence of media has become nearly impossible as it is everywhere we turn. We can't open the mailbox without finding some form of print media and/or advertisement. Taken together this media constitutes a powerful driving force. The raw power of the media is being used to program our subconscious minds.

The transhuman agenda (along with many other agendas) are cleverly inserted into *all* forms of media. This might sound ridiculous, but is absolutely true. The official movement itself is only one branch of a larger project, and the full scale of this campaign is clearly visible throughout popular culture. One of the most obvious incarnations of this is seen in Ray Kurzweil's TV appearances. Beyond this obvious example there are many more which should be obvious but are not simply because we remain unaware of the propaganda role of the media.

> The American motion picture is the greatest unconscious carrier of propaganda in the world today. It is a great distributor for ideas and opinions,
>
> -Edward Bernays, *Propaganda*, 1928

We don't realize that the primary goal of Hollywood films is to alter our perceptions. Entertainment is merely a byproduct.

We are now seeing humanized machines everywhere we turn. Think of all the movies that feature artificial intelligence, cybernetics, military super soldiers, and the like. Although these subjects are depicted as fiction, they are indeed quite real. We may not be fully aware of the military's super soldier projects, but this does not make them fiction. True science has reached a point in which it borders on the fantastic. What is presented as science fiction is quickly becoming science fact. The many recurring themes to be found in our entertainment are crafted with purpose. We are experiencing one large PR campaign to ease us into a radically different future. We are steadily being programmed to accept transformative technologies. To be truly effective, propaganda has to be fun. It needs to influence us when we least expect it. This is why entertainment has been used as a "weapon."

Would it surprise you to learn that some of the most blatant examples of transhuman themed films are actually produced with major assistance from the Pentagon? Don't be astonished, because this is a fact. The actual films will remain nameless here but it shouldn't be hard for anyone to figure out which ones are being alluded to. When you see a mass of military vehicles, slogans, and personnel on screen you should know that you are watching military propaganda. The many branches of the U.S. military, along with many other federal agencies such as the Department of Homeland Security, all have their own Entertainment Liaison Office. These work directly with major Hollywood filmmakers, television producers, and other culture creators. Not only is this done to make each particular federal department look good, but also to fulfill the strategic goals of each department. Specific themes are injected into scripts. In fact, these offices have the power to completely rewrite scripts, develop characters, or outright deny cooperation to their Hollywood counterparts. Major producers are quick to agree with the demands of the Pentagon because it means they get to use military equipment, personnel, locations, and consulting for little to no cost. At the expense of the American taxpayer, the military hands over its resources to Hollywood. In turn those same taxpayers go to the theaters to pay additional money for a movie ticket so that they may unwittingly watch military propaganda. Quite an operation, isn't it? The weapon of propaganda has found a powerful vehicle in film. The military and Hollywood work together on high budget PSY-OPS.

> The motion picture by means of its 'reality' integrates the spectator so completely that an uncommon spiritual force or psychological education is necessary to resist its pressures.
>
> -Jacques Ellul, *The Technological Society*, 1967

Our subconscious is being bombarded with visions of artificial intelligence, cybernetics, virtual reality, nanotechnology, space travel, and much more. The desire to make these things into a reality is now growing. The transhuman vision has been given to us in the form of science fiction for a very long time. Why wouldn't we want to make it real? Now that technology is converging, and these amazing things are possible, what would make us back away from creating them? Why shouldn't we accept a global alliance, which is driven by technoprogressive idealism? If implants can dramatically improve our brains, then why would we choose not to take them? If AGI could decode the universe, and bring us profound understanding of everything, then why would we not build it? What reason is there for not engaging the transhuman vision?

The reason for not continuing upon this path is that we have all been unconsciously led along by a scientific method. Our thoughts and desires have taken shape due to the efforts of social architects. Deception has been employed in an effort to mold our minds. What makes you think that these trends will change when we all decide to enhance our performance? If anything, technology will extend the power of mind control to even greater degrees of effectiveness. A computer microchip implanted directly into the brain that can beam completely fabricated reality into your mind; this would be a perfect tool for social control. Honestly, how does one prevent a dystopia in the face of such technology? Think about the words of Edward Bernays and put them into a transhuman context. All of the work that goes into creating propaganda could become unnecessary. Artificial intelligence could take over the job by constantly feeding a virtual reality into the minds of the masses. Humans could quite literally become automatons.

The sad reality of life in a democracy is that our perceptions of freedom are simply constructs of the mind. Worse than this, they are carefully crafted illusions designed to psychologically disarm us. They pacify us by instilling a false sense of security. We are told exactly what we want to hear, not what we need to know. We are told that our voice is heard, that everyone has a say in guiding politics, but the truth is that we have no say. As the scale of government continues to grow, our personal power diminishes. Agendas are planned and implemented by a socioeconomic elite. We accept these agendas because we have been bombarded with propaganda that breaks down our ability to critically think. This is the reality of the govern-*mental* process. Put quite simply, propaganda is a form of mind control. We are told that in a democracy, the majority rules, but what happens when a small minority of social technicians manipulates this majority?

Unfortunately we live in a time where propaganda does most of our thinking for us. It has become so entrenched in our daily lives that we never even notice its influence. Our minds are being manipulated with scientific precision and accordingly, we remain oblivious. The only way to counteract propaganda is to become aware of it. A conscious mind can recognize propaganda. However, this is not an automatic function of our brain; it takes serious effort to engage. There are many people who have already become so accustomed to automated devices that they expect their brains to work in the same fashion. Propaganda actually serves to promote and expand the automatic/ mechanistic brain. What a conundrum, but how appropriate. In this we see why we are moving toward a cybernetic merge of man and machine. We are merely fulfilling that which has already started. Maybe instead of debating whether or not machines can attain consciousness, we should begin directing this question toward ourselves.

> When man himself becomes a machine, he attains to the marvelous freedom of unconsciousness, the freedom of the machine itself.
>
> -Jacques Ellul, *The Technological Society*, 1967

Propaganda misleads us into patterns of behavior that we wouldn't naturally adopt. It hijacks our unique mind and replaces it with a standardized model capable of social obedience. It makes us content to live within a cold, uncaring, and inhumane system. Our standard has got to change. The mold must be broken. The only hope for us now is to realize how far we have gone and make the choice to do something about it. We must engage reality before our lives become total fantasy. We must regain our human consciousness.

-Chapter Four-
Paradise Engineering

It is upsetting to realize that the media in which we have put our faith and trust is actually being used to coerce us. The hive mind of society is programmed to believe an illusion. Sure, the illusion is "based on a true story," but to deny that propaganda permeates the airwaves is to deny reality itself, so let's get real. We have been fooled, but why do we continue believing that illusion is real? Is the media completely to blame, or is there a deeper answer? How could massive populations possibly be duped? If we were being lied to, certainly someone would notice it, right? The thing is that someone has noticed. Many people have called out the PR industry and the Pentagon-supervised mainstream media for what they truly are: one massive propaganda singularity. Instead of listening to the accusations made by such individuals, most of us have chosen to dismiss them as mere fools. The media is quick to attack such accusations and launch counter-strikes to protect its own image while simultaneously discrediting the awful truth. In most cases it is actually preferable to ignore accusations altogether. Our information is being filtered. Our ideas about the world are formulated by the information we receive. When information is censored then our ideas will be incomplete. At best our own personal view of the world will only offer a fraction of the whole, and at worst it will be skewed so strongly that it actually becomes a complete virtual reality. Either way, we are in trouble.

The truth of propaganda has been clearly defined by its practitioners. It is the valuable information delivered candidly by Edward Bernays, and other members of the "invisible government," that is used by members of the alternative media to help prove a point, that point being that we are all in a bad situation. If the accusations made by the alternative media are true, then why doesn't everyone immediately recognize the error of their ways and decide to fix things? The reason why real change doesn't occur is that it takes hard work to actually bring about. Instead of investigating the accusations made by those who question the mainstream media, it is far easier to dismiss them as fools. Their objections fall on deaf ears out of mere convenience. If we knew that something was fundamentally wrong in our world then we would be obligated to do something about it. We would be forced to action and hard work. The last thing that most people want, or need for that matter, is more work. They have been worked to death. What they want is peace. If a reassuring voice comes along telling them that they are being taken care of then all the better. Denial is no real solution to any problem, but it does grant a false sense of security. This is why so many people willingly choose to live in denial; because it is comfortable. Denial fuels our entire consensus reality. It allows a momentary feeling of peace that perpetuates itself indefinitely as each generation shirks its responsibility to confront the real truth. Confronting reality is so terrifying that most choose to die, before facing up to it.

Peace is hard to come by in this strange world of ours, but it remains a universal aspiration, a high ideal to strive for. It is what everybody wants but no one can seem to get. It is the prerequisite to establishing paradise. In ages past, paradise was promised by religion. Today, more and more people abandon traditional religion in favor of scientific humanism. The old path to paradise is being rerouted to fit this circumstance. The goal of peace on Earth and goodwill toward men has not been discarded; if anything, it has become increasingly literal. Religious paradise is traditionally thought of as a far-off spiritual destination, a heaven above whereas materialistic science concerns itself with the here and now. Science deals with what can be physically proven true. The promise of a far-off heaven achieved upon death is not enough to please someone who is of the scientific mindset. Instead of abandoning the ancient promise of eternal life, it is now being offered in the most tangible sense. Life extension technology is being developed. Experts in this field (gerontology) have claimed that an end to aging is within their grasp.

Aubrey de Grey, a prominent life extension researcher and advocate, stated in the *Journal of Evolution and Technology*,

> The support of only a small (though preferably wealthy) minority of society is required to allow the relevant science to proceed as rapidly as it can. As regards the rest of society, a muting of their opposition to such a goal is all that is needed.

This quote of course refers to the war of ideas that is inherent in bringing about life extension therapies. The actual science is secondary to this, a piece of cake if the funding is in order.

Immortality is no longer devoted solely to religion although it isn't fully divorced from it either. Paradise is now a scientific project. Advanced intelligence will supposedly allow us to decode the mysteries of nature so that we may ultimately go beyond their limits. The "miracles of science," an old Dupont chemical company slogan,

Paradise Engineering

may very well bring about everlasting life, material abundance, and spearhead a peaceful global society. This would be the new heaven for a new age, the actual incarnation of God's Kingdom on Earth, the New Jerusalem, and the mystic adage "As above, so below." This is the crossroads of science and religion that many fundamentalist thinkers remain blind to.

Perhaps you've heard this one before,

> Thy will be done, thy kingdom come, on earth as it is in heaven.
>
> -Matthew 6:10

You see, this is all the same thing. It must be understood that transhumanism is absolutely dedicated to these ends. It knowingly plays to our innate human desire for peace. Taking the peace and happiness idea to its limits eternal life is polished and given a new look. Timeless aspirations are being brought out to motivate people and get them on board with an agenda. Transhumanists have wisely employed ancient and powerful ideas to strengthen their own ideas. The means to their epic end lie within the realm of science itself.

> Transhumanists seek to make their dreams come true in this world, by relying not on supernatural powers or divine intervention but on rational thinking and empiricism, through continued scientific, technological, economic, and human development. Some of the prospects that used to be the exclusive thunder of the religious institutions, such as very long lifespan, unfading bliss, and godlike intelligence, are being discussed by transhumanists as hypothetical future engineering achievements,
>
> -*The Transhumanist FAQ*, 2003

Paradise isn't an ideal to simply wish for, but an engineering challenge. Paradise engineering is a label used by transhumanists to describe the creation of a world beyond suffering. Before co-founding the WTA with Nick Bostrom in 1998, David Pearce created an organization dedicated to paradise engineering. It was 1995 when BLTC Research was founded in Pearce's home country of Britain. BLTC stands for *Better Living Through Chemistry*, which is very similar to yet another Dupont slogan from 1935 that goes, "Better things for better living...through chemistry." Let it be known that in the 1960s, the hippie movement also appropriated Dupont's old slogan as they protested outside of chemical plants wearing buttons carrying this slogan. This was done in the spirit of comedic irony as the hippies were "tuned in" to the idea of expanding their minds through the use of chemical drugs such as LSD, but they were opposed to the pollution and corruption of the chemical industry. This was quite clever but not very wise. The hippies did not know that they were the victims of social/ psychological experimentation themselves.[32] All of that free LSD handed out on college campuses came with a price.

Just as the hippies were excited about the mind-altering qualities of chemical drugs so too has BLTC Research focused on the chemical manipulation of the brain. Of course we are given to believe that this is all a good thing. Its mission statement lays out its operation as an "ambitious global technology project," the goals of which are "to abolish the biological substrates of suffering. Not just in humans, but in all sentient life." They add that "Life on earth can be animated by gradients of ecstatic well-being beyond the bounds of normal human experience." It doesn't take long for BLTC to get into promises of world paradise.

The creation of eternal happiness could come about through the use of genetic engineering and cognitive science. Combining these sciences could bring about a new posthuman creation, a man incapable of knowing the slightest discomfort. Paradise engineering is not only explained as a scientific project, but a moral imperative by BLTC. Considering the fact that untold amounts of suffering could be ended, it would be ethically wrong to choose not to do so. This sort of reasoning is typical in transhumanist thought. The very idea that anyone would be against the creation of a beautiful new world engineered to be free of sorrow is unthinkable to most transhumanists, and therefore they expect the same attitude from other reasonable people.

The philosophy behind paradise engineering is specifically referred to as negative utilitarianism. David Pearce subscribes to the negative utilitarian school of ethics as the focus of his work is on reducing all forms of pain. The "negative" part of this label refers to that reduction. The title of *abolitionist* is also used to describe someone who hopes to eventually abolish all suffering on Earth. Ultimate freedom could be achieved by breaking away from the biological cages we now inhabit (our bodies). The transhuman concept of altering ourselves so completely that we become posthuman is seen as the way out of our prison of pain.

---

[32] Research the Tavistock Institute of Human Relations. It appears to be *the* prime center for scientific application of psychology to propaganda over the past century. Replace the word *Human* with *Public*, and you get Public Relations.

Paradise Engineering

Our biology consists of "selfish genes" and random genetic mutations. These chaotic forces ensure that pain remains constant within the current "Darwinian order." Because our genes are subject to forces, which ultimately cause pain and suffering, we should learn how they function so they can be altered to fit a more pleasant blueprint. BLTC refers to this process as the "Post-Darwinian Transition to a cruelty-free world." The violent world of material reality should be revolutionized by genetic engineering, better drugs, and new reproductive technologies so that we may all have more perfect lives. Under this imperative, posthumans would abolish any and all painful experiences that their old world ancestors were plagued with.

> Within a few generations lifelong bliss that exceeds any fantasized Christian afterlife can become the genetically-coded basis of our existence. If we want our kids to enjoy mental superhealth – emotional, intellectual, and ethical – then we can design their genetic makeup to ensure every moment of every day is a sublime revelation.
>
> -BLTC Research, *The Hedonistic Imperative: Heaven on Earth?*

Talk about idealism! This version of utopia goes beyond religious concepts of heaven by professing to be completely rational and materialistic. In a way this strange dichotomy makes paradise engineering seem even less tangible than the old religious practice of transcendence. The fact that paradise has always been offered in one form or another is a clear warning to any truly skeptical thinkers. It is also crucial to understanding the full scope of what is going on here. We are now being led to believe that paradise can actually be obtained if we follow a given set of directions. The new map to the fountain of youth contains a scientific key, but it is still the same old map.

The quest for eternal life permeates fiction and accordingly, our very reality is slowly being replaced with fantasy. Virtual reality is hitting the scene. Online virtual worlds such as *Second Life* offer a unique approach to interacting with other people online. Second Life offers the social networking of popular sites such as Facebook with the realistic 3D environments of video games. As virtual worlds become increasingly advanced, their popularity will grow beyond the scope of mere entertainment; they will also be used as tools for education and general communication.

Virtual Reality

The broad scope of future virtual realities is based on their original purpose, that of pure entertainment. The entertainment industry is often the first to develop powerful new IT technologies. In fact, some systems originally designed for military training facilities have later been used to make Hollywood films. Supercomputers have allowed the production of the special effects we have become accustomed to in movies. The trend toward super-realistic visual effects has grown exponentially alongside other converging technologies. As entertainment has become more and more realistic, it has expanded into new vehicles. The motion picture itself was once a completely new and cutting-edge technology. Film grew into a huge industry and eventually branched off into television. Television grew in complexity and branched off into areas such as cable networks, home theater, and video games. As each new subgenre of audio-visual entertainment has come about, our relationship with entertainment itself has become increasingly intimate.

In the motion picture's early days, people had to leave their homes to be entertained. With the advent of radio and television, the entertainment industry, and its financiers, were granted entrance into our homes. This was a huge step, as entertainment became a beloved part of our daily lives. Television itself took on the role of an additional family member, and it did so at an incredible scale. As television ownership quickly became standard in America, so too were ideas standardized. The same programming was beamed to each and every living room. The common family member in every home was the one which unified them all, the television itself. Its fictitious cast of characters became very real, as they were familiar to everyone. As iconic personalities, these characters and their stories were truly powerful creations. Television is yet another powerful tool to be used in standardizing the mass mind. When we look at emerging entertainment technologies we must not view them as completely new, but rather as evolved forms of what we have already grown accustomed to. We have taken much for granted.

Our connection with entertainment media has increased dramatically. Videogames are a prime example of this. They are incredibly interactive. Not only do you watch a videogame, you live it. You become the hero character and guide the progression of a story. When children (and increasingly, adults) spend hours upon hours playing videogames they end up dedicating large portions of their lives to this entertainment. In a very real sense they are living within a simulated world, a second life. The groundwork has been

Paradise Engineering

laid for ever-increasing levels of interconnectivity between digital devices and human beings.

Entertainment has slowly made its way to the forefront of human social life. It is no longer a special event reserved for occasions; it now takes up the majority of our free time. As we have spent large amounts of time immersed in the fictitious world of media, it has absolutely affected our behavior. The virtual worlds of television, film, internet, and video games constantly expand to new levels of significance as we bounce back and forth among them. This has become a part of our everyday lives. The real and the virtual are becoming very much alike, and the division between the two continues to blur.

The ideal version of our own lives is reflected in the stylized and fantastical media that surrounds us. We want to be perfect just like our favorite heroes and heroines. In pursuit of this standard we seek out ways to blend our identity with that of our role models. In terms of virtual reality, this idealized self becomes our avatar. Frequently people imitate fictional characters they see on TV (or internet, games, etc). This affects the course of their everyday lives. Like children, we imitate the behavior of our role models. We have become obsessed with chasing fantasies in a desperate attempt to make life better. It may not be fun to admit it, but this is true. Everyone is guilty of this to some extent. Many, if not most of us, have been convinced that this is actually a good thing. No one cares to know the true self anymore. This is an age of the persona, of artificial ideals. The superego runs wild in virtual worlds.

The power of entertainment media lies in its ability to make us happy. As electronic recreation has intertwined with daily routine its true power has come to be taken for granted. It comes conveniently prepackaged and is all too easy to access. We don't even have to think about it anymore. All we must do to obtain it is flip a switch. Instant gratification has been provided by electronic means. Think about a time in which you moved from one dwelling to another. Upon moving in you were most likely presented with a home lacking cable television or internet access. In this situation you could have simply left things as they were and not called the cable/ internet service company. This in fact would have been far easier and cheaper for you to do, but what did you do? You most likely called to have your local service provider install both cable and internet, but why? What compelled you to do so? You certainly could have lived without them, but you chose not to. The reason that you, and most other people, decide to make that call is because everyone has become addicted to entertainment. Sitting in a home devoid of digital entertainment has become a difficult task for many people. It evokes an unpleasant feeling. Honestly think about how you felt in this situation, in the absence of entertainment. It was uncomfortable wasn't it?

We need to be honest about our addiction, because it has a profound effect on the way we live our lives. Entertainment and other addictions have always offered a temporary escape from reality, or as Aldous Huxley put it, "a holiday" away from one's self. But what happens as this holiday (holy day) progressively becomes the entirety of our existence? A population that perceives the world around them as paradise has no interest in dissent. In this we see the obvious danger of paradise engineering. The idea is to completely destroy any remnant of sadness, ill temper, and boredom.[33] There is no room for pain of any kind in a negative utilitarian utopia. Negative emotion must be subtracted from the human genetic code. Beyond this, total bliss across all species of life on Earth has been envisioned. A euphoric super state is the endgame. The natural world can simply not provide such a paradise, so scientific engineers have been employed to get the job done. Anyone not on board with this project will eventually be regarded as a problem. You know how the old saying goes: if you're not part of the solution you're part of the problem. BLTC Research has expressed this attitude as they have said that,

> Benevolence will not just triumph; it will also be evolutionarily stable, in the games-theory sense, against 'defectors.'[34]

Combining genetic engineering with the knowledge of brain chemistry could lead to some very interesting results. The study of recreational drugs such as MDMA (ecstasy), heroin, and cocaine has been rigorous. These substances have a profound effect on human emotion but they are far from being perfect themselves. Not only are they temporary, they have many serious side effects, which negatively effect health. BLTC has taken the intensive study of the positive aspects of these drugs and directed it toward further technological development. The science of happiness is being decoded just as all other aspects of human biology are being decoded. The total understanding of human emotion requires advanced study of the brain. Our minds are being probed completely in an effort to improve upon a flawed design. The existence of discomfort is the supposed evidence that something is wrong with us. Instead of taking

---

[33] The establishment of what has been called an egosyntonic society.

[34] BLTC Research, *Wirehead Hedonism versus paradise engineering*

such discomfort as a cue to change that which is truly wrong in our lives, the negative utilitarian transhumanist seeks to eliminate pain and suffering itself. It is pain that is the enemy, not the cause of pain. In this way they are concerned with effects rather than causes. This is an extremely deep problem. To those who can see through pain to realize that it is nothing more than an indicator of a problem of greater significance, it is plain to see that the complete abolition of pain would be a cosmic error.

Profound levels of understanding have always been the way of achieving Enlightenment. Inner peace has remained the long-sought state, eluding even the most practiced of spiritual adepts. The creation of super intelligence is now being promised as an alternate route to Nirvana. By altering our biology and becoming transhuman, we would ideally be able to attain this elusive sense of clarity quite easily. Through the scientific study of mind itself, it is hoped that consciousness may be altered intelligently and rationally. Understanding this is crucial to understanding the religious aspect of transhumanism. It is a scientific religion based on the mystery of the mind itself.

The "Cyborg Buddha Project" of the IEET is the brainchild of James Hughes, George Dvorsky, and Mike La Torra. All three of these men have been practicing Buddhists. Hughes took a go at becoming a Buddhist monk in Sri Lanka for a short period in his youth. La Torra is a Zen priest, and Dvorsky is a practicing Buddhist/scientific rationalist. Their combined effort is focused on the creation of enhanced states of consciousness with emerging technologies. It is a project of paradise engineering if ever there was one.

> The three of us are launching the IEET Cyborg Buddha Project to combine our efforts and promote discussion of the impact that neuroscience and emerging neurotechnologies will have on happiness, spirituality, cognitive liberty, moral behavior and the exploration of meditational and ecstatic states of mind.
>
> -Hughes, La Torra, Dvorsky. *Cyborg Buddha Project*, IEET.org

A further development of neuroscience will provide advanced knowledge of how the brain functions. By understanding how the brain operates during "ecstatic states of mind," it could be possible to recreate them with scientific precision. Devices that affect the brain in a wide variety of ways have already been created and could be widely distributed in the near future. These could make it possible to purposely alter our mood, increase concentration, or direct our moral behavior. Ideally, technological augmentation could bring us enhanced willpower. In a way, choosing to upgrade one's self could be considered spiritual and incredibly moral. In a lecture entitled *Digital Serfs and Cyborg Buddhas,* James Hughes and Mike La Torra spoke to this directly. Hughes mentioned the fact that experiments already have been done with transcranial magnetic stimulation (TMS). These tests have used magnetic fields to stimulate electrical activity in the brain. Such experiments are very similar to the famed "God Helmet" of Michael Persinger. This device induced varied mystic experiences for people including divine revelation, meeting with space aliens, and sensations of time travel. A small minority of the God Helmet subjects claimed to be in the presence of god. All of this occurred due to the manipulation of the brain with magnetic fields. It is also worthy to note that some of the work of Dr. Jose Delgado was along similar lines to TMS. After leaving Yale in the 1970s, he went to The Autonomous University in Madrid, Spain. He invented a helmet of his own that sent electromagnetic pulses to the brain to induce a change in mood.

Hughes explained in this Cyborg Buddha presentation that the potential for malevolent use of such technology could not be ignored. A dystopian reality could emerge,

> We will have SOMA in the future. We will have the ability to have direct brain-computer interfaces that create not only control over mind but also control over all of our experience.

This will come about by the creation of a "dystopian super-heroine." Rightly, it is supposed that at least some people will become hopelessly addicted to this high. Hughes says that the real problem will be how to direct people toward constructive rather than self-destructive ends.

The *Digital Serfs and Cyborg Buddhas* presentation is one of the most interesting videos about transhumanism to be found online. Hughes briefly mentions the "Sisyphus Problem." The Greek poet Homer knew Sisyphus as the craftiest of men; Sisyphus is remembered for angering the gods (he's certainly not alone) and being set to the worst possible punishment that they could conceive of, work. More than mere work this myth has to do with slavery. Upon his death, in the underworld Sisyphus was put to the impossible task of rolling a large stone up a hill. When reaching the top of the hill the boulder always rolled back down and so his punishment had become eternal labor. The Sisyphus Problem, as explained by Hughes, is this: how do you actually get Sisyphus to enjoy his punishment? The answer to this problem holds great appeal to slaves and slave owners alike. Slaves have a vested interest because the key to their happiness is involved. Slave owners want to solve the Sisyphus Problem to obtain the perfect laborer.

The dark side of emerging technologies is explored in depth within this video, and it is refreshing to see real questions and moral outrage coming from the audience. Of course the IEET believes that ultimately these technologies will be used for beneficial purposes and that there are no moral absolutes. The question of whether or not it is morally acceptable to program a human/ machine to love its own slavery is left "open." La Torra says this, and he does so from an *enlightened* point of view. There is no high moral code being followed by science or esoteric religion. The reality of our world transcends our own unique ideas. Moral relativity does exist, and accordingly our own morality shall always be what we make of it. This is one of the many mystic powers of the mind. It is well known that "moral uplift" shall become anything that we want it to be. So the true question to ask is: what do we want?

Thanks to emerging technologies the moral uplift of humanity could be as easy as turning a knob. This could effectively reduce all of our unwanted destructive behaviors. We could also become increasingly spiritual. La Torra explains that the ancient path to Enlightenment has never been popular. The reason for this is that it has remained extremely difficult. Most people give up when they reach the "Dark Night" phase. With new mind-controlling technologies, the difficult task of achieving Enlightenment will be far easier. This will lead the masses to new spiritual heights. For the first time ever, the majority of the population will be lifted up out of darkness into the light of a new day. Someone is extremely interested in bringing this great gift to the world. The identity of this someone remains a mystery, but the nature of his value system is what influences advocates of the Singularity to adopt his particular belief structure.

It is critical that transhumanists make it known that their efforts toward paradise engineering are understood first and foremost as legitimate scientific projects. They are honest about deleting suffering and achieving happiness, but they certainly don't want to be classified as another religion. They want to achieve their version of paradise through the application of science. The various mysteries of spiritual experience are systematically being broken down. What once were considered mystical experiences are now being explained as simple electro-chemical functions of the brain. Science seeks the source-code that creates all of these spiritual phenomena. Mystery is systematically being removed from the human experience in the quest to achieve ultimate understanding. Old belief systems are being broken down as new systems of logical thought are being created to serve as the foundation of a new global faith.

Another online group for transhumanists known as the "Trans-Spirit List" has described the spiritual side of the transhuman project,

> It seeks to understand religion and spirituality in terms of cognitive science and evolutionary psychology, and to project the future of religion and spirituality in the dawning transhuman era.[35]

The Dawn is coming, or so they say...

Again, this is all about understanding; not a mere general understanding, but a total understanding of our material condition. A desire to know the truth can quickly become connected with a desire to control everything and everyone. The true nature of the will toward power is something that cannot be underestimated. As our level of understanding increases, so too does our ability to manipulate the world around us. To varying degrees this is what all science has been about. Many people are convinced that this is a good thing, but the subtleties continue to expand. The technological buildup of power has steadily progressed without informed oversight from the population at large. The fact that most people have been out to lunch during this process is good for those who are already in power because it ensures that their own will has no real opposition. It is absolutely necessary for these powerful individuals to remain hidden from the public eye. The best way to do this is to create diversions, which will distract people, as well as confuse them. This is why most people are ignorant of the true power and planning behind converging technologies. They haven't been paying attention and they have never thought to.

The profound level of technological power that now exists is far beyond any single person's ability to fully understand. If we continue to march forward as we are doing now, then there will be unforeseeable consequences. The creation of a new scientific religion will complement the awe-inspiring birth of superintelligence. It seems as though AGI will be the new god for a new age. Our new god will promise us the same things that our old world religions promised. Eternal paradise will be offered, if for no other reason than to create the illusion that we are all being taken care of. The paradise to be engineered could be logically created with advanced knowledge of how each and every individual human mind works. The ability to

---

[35] Trans-Spirit, http://www.groups.yahoo.com/group/Trans-Spirit/

Paradise Engineering

keep people happy for eternity may just become possible. A machine could solve the Sisyphus Problem. We have to understand that any logical approach to a problem will beget a logical answer. An AI god situation seems perfectly logical from where we stand today.

The pursuit of happiness has led us to develop new and improved methods of feeling good. The creation of recreational drugs has advanced greatly in recent history as laboratories have brought us more and more ways of pleasuring ourselves. The widespread proliferation of these drugs has been complimented by research into their many effects. Unbeknownst to most of us, these drugs were unleashed upon the world as a massive scientific experiment. Neuroscience is used to decode the cause and effect of these chemicals. Social science concerns itself with the understanding of collective human behavior, and it complements development in neuroscience. It is social science, which is the common denominator among all fields of human scientific study, because it puts all discoveries into a useful context. It is the means of deploying advanced technology on humans to kick-start the "transhuman era."

Complete scientific understanding doesn't happen overnight. It requires repeated experimentation. Organizations such as BLTC Research meticulously analyze vast amounts of data gathered on drugs like MDMA. They hope to see the emergence of more efficient drugs and technologies that will enable increased results. The power involved in creating artificial forms of bliss cannot be ignored.

Consider the effect that drug addiction has on society today. In the pursuit of momentary pleasure, drug addicts suffer many long-term consequences. The intense drive to feel good is fueled by many different factors, one of which is, of course, chemical addiction. A scientifically minded person may be inclined to believe that by engineering out the genetic predisposition toward addiction, the problem could be solved. You could eliminate the biological basis of addiction within the human body, as well as create new and improved drugs, which aren't chemically addictive. This completely physical solution to the problem could theoretically make it safe for us to create a virtual paradise in which we never feel the slightest unease. Without the nasty problem of addiction we will be liberated and can feel good about feeling good all the time. This view ignores addiction in its deeper sense.

People use drugs not only to feel good, but also to escape from reality, to avoid feeling bad. This world is filled with many unpleasant things. These rightfully cause us discomfort. By not dealing with feelings of unease directly we are actually doing ourselves a great disservice. Easy answers are not real answers. However, it is always easier to escape to another world than to deal with the real. Drugs create a parallel universe existing wholly within a tiny material portion of the greater world, a single human skull. The reality of this entire situation eludes the drug addict as he seeks to escape from the horrors of true life, but the sad fact of the matter is that there is no such escape. As the drug user experiences his artificial world of ecstasy, the real world continues functioning all around him. It remains unchanged by his state of mind. Upon returning to reality, the addict becomes increasingly depressed because he has to make up for all of the real world problems which were created in his absence. Instead of choosing to create within the real world, he consistently escapes to the virtual reality that his addiction provides. This is part of the broader existential problem of addiction. This problem will not only persist but also expand under the transhuman project of paradise engineering.

Sure, paradise engineering could provide a way for us to feel good all the time, but what does this mean? What will happen to the greater reality in which this paradise is created? By engineering constant bliss will we be shirking our responsibility to solve our real problems? Who will create this virtual paradise for us, and why should we trust them with such fantastic power? Our natural human instincts provide us with a healthy sense of distrust toward the notion of a virtual paradise. However, if a virtual paradise is indeed created, then it will undoubtedly attract many people. It will be impossible to resist because as long as one gets a taste of bliss, then they will have a difficult time turning back. It won't matter what is right or wrong once the addiction has begun. This is how addiction works. It is not foolish to recognize a profound darkness in it. Something *trans*cendent can be seen here, and it isn't pretty.

Just how addictive will be virtual reality, "mood brightening" drugs, and other forms of transhuman bliss? An even better initial question would have to be: how addictive are they already? We have become addicted to constant entertainment as evidenced by the massive amount of time that we spend watching television. Is it any surprise that we choose to spend such large portions of our lives being entertained? We do it because it feels good. Entertainment offers an escape from our own lives. It provides the very same thing that recreational drugs do. Escaping from reality offers temporary relief to a permanent problem. This is the fundamental problem that underlies the entire issue of paradise engineering.

Paradise Engineering

We want to escape to paradise because we are not satisfied with who we are. We have actually become afraid of the possibility that lies dormant within us. The ability to positively affect the real world around us is strangely terrifying to many people. Our innate power to do good deeds comes with serious responsibility. This responsibility is what causes many people to run away from themselves and seek refuge in empty pleasures. There is a profound difference between simply feeling good and doing good. When we do a good thing it benefits other people. It makes a difference in the real world. We are all affected by each other's actions and inactions.

In many ways virtual reality offers a greater sense of human togetherness. The promise of an increasingly connected world provides the vision of a better reality. There will be no more reason to feel any distress in this new world because it will be a paradise. We are supposed to believe that this is true, that our eternal search for paradise will finally be achieved through science. Posthumans will have the ability to stimulate the pleasure centers of their own minds to such an intense degree that even the thought of discomfort will be impossible. World peace will be achieved by producing a population that is totally content with the world around them. Perception will be easily altered and, of course, perception is reality. Our reality will become the product of scientific engineering. The engineers of this new system will have elevated themselves to the status of gods. Does this sound like a good idea to you?

-Chapter Five-
Total Destruction of Privacy

A big problem with paradise engineering is that everyone has his own unique idea of what paradise should be. World peace would require the reconciliation of separate governments, cultures, and peoples. The spread of mass media and communications has done much to speed up the process of global consolidation by standardizing culture on a whole new scale, but it cannot do everything. Individual humans all over the world still hold on tightly to their traditions, and they all have their own unique ideas of what paradise should be. Most people's version of paradise involves attainment of peace and quiet. The average person wants to be left alone, to be able to live their own life without the intervention of hostile forces. Indeed, most people are completely happy and content to live the simple life. The specific details involved in making this ideal a reality vary from person to person, but the fundamental yearning for this peace is undeniable.

The problem in attaining peace and quiet lies in the fact that a return to the good old days is becoming less and less of a realistic goal. The modern world is becoming increasingly dependent upon technology. Increased technological power is directly linked to the consolidation of world cultural, social, governmental and economic systems. Technology has enabled progress toward world unification. There has been a deliberate effort to use it in this way.

The idea of world peace has been directly connected with the creation of a one-world system. There is a definite reason that this idea is being propagated and that is because a world governmental system has been building up for a very long time. It is slowly making itself apparent, first through subconscious suggestion, and later through more overt media cheerleading and public acceptance. We have all been subject to this campaign and as such we are now at a point in which we struggle to make sense of what is actually happening all around us. The mass mind understands the socially engineered version of world peace, but individual minds still hold on to their own version of the good old days. This conflict within the individual creates severe confusion over identity. As the good old days rapidly fade into the past, it becomes increasingly difficult to know what to do. For most people, cognitive dissonance takes over. In the face of constant change and uncertainty, most people become so confused that the only thing they are able to do is go with the flow. Accepting change occurs by default due to the lack of any other viable options. Anyone wanting to steer massive change in a particular direction would do well to make the population at large feel as overwhelmed as possible. Cognitive dissonance leads to apathy. An apathetic population will accept anything.

You must know that there is a definite reason that the generic idea of *change* has been so heavily promoted throughout all media. This has occurred because a huge change is being planned and will come. When it does come, the mass mind will be primed and ready for change. The exact nature of this change is what we must now understand.

Propaganda has encouraged us to *hope*[36] for a better day in which peace can finally reign over the Earth, but before we get too wrapped up in blind hope we should take a moment to look at the specific trends at work in the world today. Any sort of change made will be a direct outcome of what we are doing right now. If we are ignorant of the truth of our situation then we will never understand what is actually happening, or what is to come. The true change, which has been occurring all around us our entire lives, involves the global consolidation of governments, economies, and peoples. Beyond establishing a new world state, there is also a plan for the genetic reconstruction of the human race itself. Because both humans and machines have served as tools in the creation of this all-encompassing system they will in turn be perfected by being adapted to the system itself. The world is becoming a unit and all of its sub-units are being upgraded so that they may run more smoothly and efficiently. This is the truth of the Singularity, and we can clearly see it is also what the transhuman merge of man and machine is about. The end result is ultimate union, created not only by bringing together biology and technology, but the entire world itself. This project is well underway and is being implemented in the name of one common cause. This is the truth behind all of the touchy-feely propaganda you have been spoon-fed your entire life. The good old days, and the individual human now face extinction. Hope is for dopes, and as intelligent individuals we need to drop it. Hope directs your attention to an exterior object. It gives you an idol to worship as it turns you away from your own center. This is how we have all been disarmed.

---

[36] Our current president ran on a campaign that used both of these powerful buzzwords: *change*, and *hope*.

Total Destruction of Privacy

The role of government in regard to convergence is to oversee and accommodate its actual process. As the new system progresses it requires new policies to accommodate its further development. This process has been ongoing for a long time now. To believe that government has any interest in preserving your old way of life is naivety of the highest degree. The good old days are precious to the individual, but a hindrance to change. Rapid technological innovation is indeed hindered by old systems. This is especially true of the American constitutional form of government, which focuses on the individual citizen's rights as a free and sovereign being. This is the antithesis of the emerging global system. In this system the individual would only be important because he is part of a larger unit, an overarching and interdependent singularity.

New inventions are created all the time that cannot be used because of their violation of legal human rights. The push for convergence involves many such technologies that pose problems for our civil liberties. A developmental halt has not occurred. However, the actual application of technology is stifled because of this. This angers its creators a great deal. Utopia is being held up because of people's outdated ideas about themselves and their rights, but don't think that this holy project has been cancelled. It is without a doubt gaining momentum. Day by day, brick by brick, the temple is being built. We have already seen a massive change in our legal system because of this.

Human rights are being systematically reduced as governmental power increases. We are supposed to believe that this is all part of a natural process. We are not supposed to question change, we are expected to simply hope for the best. The progress meme has done much to break down our collective human willpower. There is no doubt that we are on the edge of dramatic change, but what does this actually mean? What lies beyond the fog of ambiguity? Is it the passive human acceptance of a global singularity and a posthuman future?

In this "post 9/11 world" it has almost become cliché to mention the extreme disintegration of American civil liberties that has taken place in the name of the War on Terror, but if we are to understand the reality of our situation then we must do so. Our constitutional rights have been obliterated in the wake of 9/11, of this there can be no doubt. We have been told time and again that we need increased security to keep us safe from terrorism. But how secure are we in a world where we have given up everything that truly matters? Our basic human dignity depends on upholding our individual freedoms.

As our rights are whittled away one by one, we in turn come closer to an unseen end game. This is something that transcends all political issues, for it merges them all. The left/right dialectic of American politics is a diversion of the cruelest intent for it not only limits people's vision, it also pits average people against one another. In this way we destroy ourselves as everything of true value is stripped away.

It is incredibly hard to obtain freedom of any kind in this world. Once you give up your rights, they are incredibly hard to win back. That is why we must understand that the systematic deconstruction of our constitution has been deliberate. Anyone telling you otherwise is either a fool or a charlatan. There are plenty of excuses as to why we have let this happen, but none hold water. The preservation of individual human dignity has no price.

The fact that the NSF/ DOC workshop on NBIC technologies occurred less than two months after 9/11 went completely unnoticed by us. Our attention was immovably fixed upon the enemy who had attacked our homeland. The immediate danger of terrorism was the only current event of relevance at that time. While Americans were stuck in their terror-induced tunnel vision, the professionals at this conference were free to discuss their own long-term goals. The power of converging technologies was no mystery to government and big business. This conference was but one of many operations designed to plan the developmental strategies for rapid scientific growth. Appropriately enough they discussed future technologies that could be used to prevent terror before it actually happens. The complete monitoring of everything and everyone would allow powerful AI to accurately predict the future in order to prevent any occurrence of violence. This sort of societal control would require total surveillance of everything and everyone. There is no way that such technology could have been used within the confines of our previously existing legal system. This is why laws have since been altered.

There is a reason why the PATRIOT Act was signed into law less than a month after 9/11. It was part of a massive power grab on the part of the federal government. The legal process of destroying civil liberties was a precursor to the emergence of advanced technological control mechanisms. The level of sophistication of Intelligence Technology within the Intelligence agencies has continued to remain unknown to us. This is a simple matter of national security.

The creation of massive databases and the subsequent analytics and pattern mining of their information had to be legalized. Before 9/11, there was no precedent to totally monitor American's telephone, internet, and financial records. Doing so would have been met with outrage from the public. The PATRIOT Act arrived at a time in which people could accept such a change in policy. Since that time databases containing citizens' private information have been created in the name of fighting terror and there have been many other pieces of legislation passed that have increased the government's power to collect information on the citizenry. Many stories in the media have focused on the NSA's Terror Watch List and other such programs that have concentrated on the creation of large lists of Americans to be monitored due to their possible connections with Al Qaeda. These stories often throw out specific numbers showing the number of people who have been added to such lists. In 2006 the NSA publicly announced that they had stored the telephone call logs of 200 million Americans. A crucial point to be made here is that this figure is not necessarily completely accurate. It is impossible to verify these numbers and it is likely that they are far greater than reported. This is simply because the NSA has the right to withhold such information in the name of national security (this is the claim at any rate). If the NSA were keeping the records of every single telephone call of every United States citizen, then legally they would not have to let us know.

A secret program approved by the President in the fall of 2001 gave intelligence agencies the ability to completely overlook the fourth amendment and bypass the FISA court system. All this amounts to in the grand scheme of things is an upgrade to the efficiency of domestic intelligence operations. Complete and total knowledge must be obtained over the population. Programs such as the Pentagon's Total Information Awareness (TIA) project of 2002 have sprung up as a result of this. This DARPA project was aimed at advancing data mining technology. It worked toward a more complete merge of commercial and governmental databases. It was sold on the premise of searching for terrorists globally, but it absolutely enabled advanced domestic spying. Admittedly, this project set out to compile the email, calling records, credit card/ bank transactions, and travel documents of American citizens. This would enable the creation of detailed profiles on anyone within the U.S. Not only intelligence officials, but also military and law enforcement would be able to use TIA technology in a collaborative effort. The federalization of local police forces has only continued to be carried out since TIA was officially abandoned in 2003. It is appropriate that TIA officially came and went in less than two years. This at least gives the public the impression that the project was abandoned. Don't count on it. The DoD has continued serious data mining projects since that time.

Government now has the legal right to work with communications service providers to further its data mining capabilities. The 2007 Protect America Act was another piece of legislation specifically drafted to further this task. The official line given to the press by the presidential administration was that the act was needed to close a "surveillance gap." The international scope of the war on terror has been used to convince us that compiling such lists is a necessary security precaution. Anyone who has somehow made it onto one of these lists may have ties to radical terrorist groups abroad. We have been told that this is a way of routing out domestic "links to al Qaeda and their known affiliates." Two hundred million people on a watch list is meant to sound reasonable in the context of National Security. The explanation that this is being done solely to find terrorists is misleading. We have to see the broader scope of these trends and understand what they are enabling. A massive surveillance system is now in place and is being used on the entire American population. To many this sounds like a bold statement, but when you analyze the big picture it seems all too obvious. This massive national security system has been created under the guise of keeping us safe. Bit by bit we are being psychologically conditioned to believe that we need a total surveillance system in order to survive. It may take a while for people to completely accept such a thing, but until then the work will continue to be carried out covertly. The worst part is that this is all becoming blatantly obvious and we continue to do nothing about it.

As the nationwide, and worldwide, data collection trend has furthered, so too has the secrecy of government. The 2002 Homeland Security Act had specific provisions which made the newly created Department of Homeland Security immune to Freedom of Information Act requests. It was this act that actually enabled the Information Awareness Office (IAO), the office that carried out the TIA project. The War on Terror has been covert. Its very identity is tied to our National Security agencies. We have already reached a point in which nearly all of our personal information can be obtained secretly. Many excuses have been made to cover for this massive destruction of privacy, but the fact remains that this is all being deliberately carried out to create a new, more efficient management system for society. The frail guise of fighting terror has been, to a great extent, rejected by a large portion of the population, however the trends of increasing governmental secrecy and massive data collection are converging into something far greater than even the most paranoid among us have imagined. Our knowledge of technological development and integration is lacking. We lack the

Total Destruction of Privacy

vision,[37] of those who are invited by the big government departments to plan such things.

The legalized surveillance of the American people is being combined with the power of massive new IT systems. These systems are being built to decode the complex system that is society. Humans are regarded as complex systems themselves. In fact they are complex systems made of many smaller complex systems, i.e. the nervous system, digestive system, reproductive system, and so on. All of these things are subject to scientific study in order that the transhuman project of genetic engineering may be perfected. The decoding of biological data was first seen on a large scale with the Human Genome Project (HGP). The DOC and other federal departments have brought the HGP and many subsequent bioinformatics projects to you. This has of course been facilitated by corresponding projects in artificial intelligence. The many and varied complex systems that make up the whole of humanity require complex tools for data collection before any changes can be made to them. Beyond merely collecting data, AI pattern mining systems are needed to make sense of this ridiculously large amount of information.

Converging technology facilitates the decoding of complex systems at all levels. From the secret operations of the federal government, all the way down to the individual consumer, there is a digital monitoring device in place for all occasions. Huge computer systems designed to store and monitor everyone's activities are useless without effective pattern recognition systems. With the advent of more powerful AI, this need will be fulfilled. Our current digital world provides the initial framework for an emerging global surveillance system. Take a moment to think about how many of your daily activities generate a digital record of some kind. Credit card purchases, cellular phone calls, and all internet activity are prime examples of this.

The trend toward increasingly powerful digital technology has a dark side which most people don't feel comfortable admitting to. We don't think twice about what we do during our daily routine. We are handed cards to pay with and smartphones to play with, and we just love it. Little do we realize that the myriad of electronic toys we have come to know and love can actually be used to monitor us. Anyone who chooses to deny this fact of modern life does so at his own expense. Laws, which have legalized the process of spying on the entire population, were not made to simply find Americans who are

---

[37] The NBIC Report uses this very terminology.

linked to international terrorist networks. These laws were passed to spy on everyone. All of the computers and personal digital devices that have exploded onto the consumer market during this time are serving a dual purpose. These machines, which we naively believe are our friends, are actually double agents. They serve us by offering increased communication and understanding of the world around us while at the same time serving the global surveillance system by monitoring our every move. Common sense tells us that privacy-destroying laws can be abused. When we read the various policy reports on converging technology it becomes very clear that the desire to completely monitor us is common among the governing class.

Look at the history of the Global Positioning System (GPS) and you see that it ties in with aerospace programs. Just after Sputnik launched into orbit, satellite-based navigation systems were tested by the U.S. Navy. "Transit" debuted in 1964, "Timation" in 1973, and also during that year the Navy began working with the Air Force on their NAVSTAR program. What started as a military project has since integrated with the private sector and law enforcement. Satellite-enabled tracking systems are used for many purposes now. Onboard vehicle systems are both marketed to consumers and used by companies to track their fleets. GPS has already been used in autonomous vehicles (vehicles that drive themselves) and could be used to a far greater extent. An entire autonomous highway could actually be built with today's technology.

If you remember, in 2005 the federal e911 program became effective. This was an FCC mandate requiring all cellular phones be equipped with technology enabling them to be tracked within 100 meters. This was done to ensure that emergency 911 calls could be tracked to a specific location. GPS wasn't needed to make this program effective, and its advent has made tracking even easier.

GPS is now built into many cellular telephones. Tracking people is one of the main reasons that this has occurred. The cellular service providers have the ability to track all phones attached to their networks. Wireless networks, wi-fi, and internet navigation services can now be used to track phones. Earth's surface has been systematically mapped out. Linking phones to digital maps allows the tracking of an individual's movement. Not only can service providers (who are required by law to accurately track you) do this, average citizens may now use various services to track the moves of others. Today we are even hearing radio advertisements attempting to sell cellular tracking services to us. People are being encouraged to track their family, friends, and employees.

We have now acclimated to the fact that cellular telephones can be tracked and traced. This fact has made its way into the themes of many Hollywood films. The characters become paranoid about being found so they decide to destroy their phones. This is a friendly reminder from your culture creators to you that your phone can indeed be tracked.

Understand that any system with the power to record the position of a phone in real time could be coupled with an I.T. databank to permanently store a record of when and where that phone has traveled. Obviously, this could provide a permanent record of the cellular phone owner's movements. Such systems have actually been proposed in the context of catching speeding drivers. When cellular companies are legally bound to give your phone records to the federal government upon request, what follows? It seems clear that communication service providers have already become a defacto arm of the federal government. What do we think happens when such broad powers are given to government? How can corruption be avoided with these provisions in place? How much of our daily activity is being recorded, data-mined, and permanently stored? When such operations are carried out secretly, we simply don't know. This is a serious problem.

Beyond tracking our movements with cell phones we can also be tracked by our purchases. When you use a credit card to pay for something the transaction is logged digitally by your bank. Not only is the dollar amount recorded, the exact location and time of the sale is too. Again, it is appropriate to point out a popular movie reference that we have all become familiar with. The criminal is always sure not to leave a paper trail of his sordid financial activities. In doing so, he is able to get away with murder. Decent, law-abiding citizens have "nothing to hide," and so they don't think twice about using their credit cards. The electronic payment system has now become a familiar part of our daily lives, and as such, we have taken it for granted. The money that we use continues to become increasingly virtual as commerce is carried out via computers. Online shopping is a prime example of this. The entire process of buying and selling is going digital.

Our internet activity is extremely traceable. With huge international surveillance systems monitoring internet it should come as no surprise that online activity provides huge opportunities for data mining. ECHELON is one of the well-known systems that uses advanced filtering methods to track individual people. In this we again see the handiwork of the NSA, that same agency which has been busy compiling the phone records of at least 200 million Americans. For this project they worked in tandem with other nations' security agencies, because government is now global. Not only can keywords be filtered out by powerful systems such as ECHELON, but also the complete content of every webpage that you visit can be recorded. Systems with the ability to scan through this information and detect patterns can extrapolate detailed information on individuals. Through the use of algorithms, these computers can understand the subtle nuances contained within the patterns of your internet activity and in theory, your very mind. They can detect patterns within patterns to develop a detailed personality profile on you. These massive systems are psychoanalyzing us all. This is what happens when our lives are catalogued digitally. Artificial Intelligence has been created for these purposes. Its very role is that of intelligence and it is only natural that it is used by intelligence agencies.

By mixing cognitive neuroscience with powerful information technology, it is possible to recreate our real world within a virtual simulation. This is what is known as "realistic agent-based social simulations." The "agents" represent individual human beings, groups, institutions, etc., and the corresponding "social simulations" are digitized versions of the larger society in which these agents exist. Given the fact that our daily lives are now digitally monitored, it is possible to use such simulations to forecast the future. By creating virtual simulations of real people it may be possible to predict their future actions. Information collected from digital surveillance systems makes prediction more accurate. Huge pools of data are useless on their own. They need to be put into context with these simulators. The Pentagon's Sentient World Simulation is likely one of the most advanced social simulators in existence. It would only make sense that this is the case because most emerging technologies are created by the military in the first place. After they are developed and put in place for strategic/tactical use, corresponding consumer-level versions of the very same technologies may then be given to the public. Consumer products are enabling a worldwide surveillance system.

The intense scrutiny of our daily lives has a definite purpose, and that purpose is ultimately to control our behavior. Of course, you can't control what you don't understand and this is why every effort is now being made to understand the totality of human behavior. By digitally recording as much information as possible, an ocean of patterns may be fished by powerful AI systems. This is happening and as it does, our behavior continues to be altered by technology itself.

Our material and virtual lives are converging. The many "social networking" sites that we take for granted are actually aiding in the creation of social simulations. Social networking takes our daily activities and turns them into digital records. We are actually choosing to spend increasing amounts of time in virtual worlds. Social networking websites such as Facebook, Myspace, and Twitter have become a phenomenon. They quite obviously mix the real and virtual worlds together, and in the process they create an entirely new social dynamic. Everyday life is beginning to transform due to our constant digital connection. Information is instantly accessible on all persons who choose to "update their status" on one, two, or all of the major social-networking websites. This is a power which is being exploited. While serious digital psychoanalysis is done upon us, we go about our social networking without a care in the world. Surveillance has come in a form which appeals to us. Unlike the secret operations of the NSA, the addictive quality of these alternate realities is quite obvious. We have reached a point where online discourse has become ridiculous, with people sending out updates such as these,

> Scrambled eggs and lobster so good
>
> oh what a glamorous life. just unclogged three drains.......
>
> going for a run
>
> never been this hungover buying this much beer :/
>
> So I'm an idiot & since I was in a rush I forgot to wear a longer tanktop or something under the shirt I have on & now my ass is hanging out

I literally collected this set of "tweets" within a span of five minutes. The actual twits behind the tweeting will remain anonymous, however it should be noted that on the actual website, these messages are labeled with the time, date, location, and name of the person posting them. At first glance this all seems so trivial and unassuming. It is all too easy to cast these strange examples of modern life aside. Once again, the bigger picture eludes us when we decide to take these things at face value. We need to understand that social networking provides an incredibly powerful tool for information gathering. All of this data is perfect for tracking trends and indeed, it is used for that very purpose. There are many websites that monitor Twitter. On these sites you can see what people are talking about and where they are located. Actual maps are generated with trends highlighted for each and every area of the world in which people are tweeting. The most popular words let you know what the hive mind is interested in at any given moment. A search just after Halloween showed that the word *costume* was very popular. This of course is pattern-recognition at work, something that AI developers are very keen on. Not just AI, but market research and intelligence gathering can be done using tools such as these. Given the sophistication of public websites that track twitter, one has to seriously wonder about the level of technological advancement that the NSA must have in this same arena. How long has it been tracking internet activity like this? Are these tweets being databased? What patterns are being discovered within this gigantic amount of information? Are AI bots being put to the task of decoding an internet hive mind?

In 2010, internet giant Google openly admitted to working in tandem with the NSA. Given the fact that this company has long stated that it wishes to compile all of the world's information into its digital databanks, we have some serious food for thought. The founders of Google have expressed interest in building an AI system powerful enough that it could be accurately described as the "mind of God." This is how they have described their vision. They use the "god" term in much the same fashion as transhumanists: to describe an all-powerful machine mind. Note that Google is consistently at the forefront of privacy-invading services. Its Latitude service was one of the first to offer cell phone tracking. They go around the country in their vans taking photos of every street in the country. They openly admit to monitoring the microphones of people's personal computers. Every chance it gets, Google collects more and more digital data. Combine its already successful digitization campaign with advanced AI and you have something to think about. A federal/corporate singularity has been openly announced. In 2010 Google admitted to working directly with the NSA. One must wonder just how long this partnership has been active. Again, it is very likely that this has been the case since Google's inception. Powerful tools for massive data collection are already in place and they have assumed the identity of free services. Don't let such labels mislead you; there is a price to pay for these fun and convenient products.

Not only do people post huge volumes of information about their personal lives, pictures, favorite books and movies, physical attributes, mood, and more, online, but they also create connections to other persons by becoming online "friends." This creates a digital web of connections between real people. Most individuals use social networking sites to stay in contact with the people that they meet in their real lives, i.e. their actual friends. When real world friends become online friends they become nodes. Just another connection in a sea of connections that makes up the world wide web. Our personal lives officially become public knowledge. One doesn't even need to covertly spy on a person who posts this information willingly. It is easy to find out who knows whom, and understand why those people may have become friends in the first place based on their common interests. AI bots and federal spooks aren't the only ones who can do this. Ordinary people now have the freedom to pour over these websites and get a glimpse of various social networks from across the globe. Considering that any idiot with an internet connection can learn all about who you are, just imagine what could be done with powerful data-mining tools.

It is no surprise that the market research power of these technologies is being exploited to the utmost. The power of such internet data collection has already been prominently displayed on the social networking websites themselves. They have featured advertisements specifically targeting users based on their identity, i.e. profile settings and web activity. If you spend your time online searching for organic watermelon seeds, then an advertisement for organic watermelon seeds will pop up on the site. If your profile displays that you are a heterosexual male, then you will receive advertisements that promise to introduce you to women in your specific town, city, or county. These targeted advertisements are only a glimpse of what is possible with the power of *total* information awareness.

It is important to understand that social networking technologies are being used by young people far more than by their elders. The youngest generations are being born into a world of constant communication. They don't question this because they believe it to be completely normal. They have never known a world without personal computers, and so they take their situation for granted. We can see in this example an eternal truth. Most people believe that the world surrounding them that shapes their very consciousness is totally natural. They don't spend adequate time questioning their own situation. This is what allows constant change to occur, but what is guiding this process?

A completely transparent society is emerging: a world where complete surveillance is not only accepted, but welcomed. Ironically, total surveillance is actually being sold to us as the solution to all of our security issues. The very entities that have ruined our individual security are now marketing their giant technical apparatus as the solution. Abandoning these technologies or taking some alternate course is not given as an option. Instead we are to become even more digitally connected as government rewrites the rules that control our society.

AGI researcher, top transhumanist, and Singularity University's director of research, Ben Goertzel has consistently advocated "sousveillance," which is basically the complete eradication of privacy in favor of complete information sharing. Under sousveillance everyone has total access to all information, and thus people watch each other's every movement. In March of 2009, at the second Conference on Artificial General Intelligence, Goertzel gave a lecture entitled *Is Sousveillance the Best Path to Ethical AGI?* This lecture is available to view as an online video and it is highly recommended that you do. During this lecture Goertzel claims that privacy is now impossible and that sousveillance is better anyway. He mentions the fact that intelligence agencies are collecting huge amounts of data on us now. With a smirk he says that Google databases, credit card transactions, and emails are, "most likely being piped to other repositories run by governments... why aren't they doing more with it?" The answer to this question is that mining patterns from these databases is incredibly hard. He argues that this is why powerful AI/ AGI must be created. Goertzel mentions that sousveillance would encourage conformity, which in turn would lead to a very predictable society. Again, we see the concept of total predictability popping up.

Portable phones have become powerful computers. This has revolutionized the entire human social dynamic. Not only adults, but now also children have their own personal cellular phones. Youths born with this technology in their hands learn to use it in the most impressive ways. Text messaging has become second nature to the children of the information age.[38] Social networking is now creating an augmented reality filled with internet, phone, and real life communication. This amalgam has already become the norm. Being disconnected from the electronic network is torturous to those who have become addicted to it. Parents now punish their children by taking away their cellular phones or disconnecting their internet. This speaks volumes. We could learn much about what is happening to us if we had the strength to observe our own behavior impartially. Most

---

[38] Actual text messaging tournaments are now sponsored by wireless companies. The winners have consistently been young teenagers.

people are well aware of the fact that addicts are most often completely oblivious to their own addiction. Technologies are in fact creating addictions and we are none the wiser. If we take the time to see that mere objects, which are external to ourselves, are creating such powerful addictions, we must also be humbled. Our addiction is nothing to underestimate. Just imagine what would happen if similar technologies actually became one with our biology. What sort of addiction would a cyborg be prone to? Would enhancement enable extreme addiction? Is addiction being willfully exploited already? If so, then why would this trend not continue? What makes the most sense, practically speaking? Understand the cold, calculating, and inhumane quality of practicality, and you will come face to face with the awful truth.

Human beings are becoming machines. This is being done, quite simply, to make the whole of humanity more efficient. This point is specifically mentioned in government reports, and yet we don't fully understand what is happening. The problem is that we aren't reading these valuable sources of information, and we truly aren't expected to. What is expected of us is to go about our lives as if nothing out of the ordinary is happening. A virtual prison is being built all around us, and it remains invisible as we go about living our lives, which on their own could be described as virtual realities. Our real-life behavior is based on false information. Instead of questioning our strange predicament, we choose to normalize it. We take what we are given because asking questions is a hindrance to anyone who simply wants peace and quiet. The reality is that ultimate peace and quiet is indeed part of the plan. Although, the nature of this peace is not what we all have in mind individually. The global version of peace and quiet is being instituted slowly but surely and it relies heavily on technology. In the end technology will be tasked with maintaining order. This it will do methodically and efficiently. All obstacles to such efficiency must be eliminated.

As Aldous Huxley said at Berkeley in 1962, all "inconvenient human differences" must be "ironed out" if the "Ultimate Revolution" is to occur. This revolution concerns itself with the mind, for it is the mind that animates the body. In the new age the body social is to be animated by a new way of thinking; in fact a new mechanical overmind of massive proportion may direct it. Technology will be integrated with biology to maintain a peaceful new way of life. People are being turned into things, objects to eventually be used as the automated systems of tomorrow. Many like to talk about how machines will one day do all of our labor for us without ever stopping to realize that these utopian machines will actually be our descendents. This is the fate we are haplessly leaving to our children as we never take the time to understand what is truly happening in this world of ours. We have never chosen to understand what slavery actually is. In our self-righteous quest for progress, we haven't ever admitted that we are all slaves now. Abolition of slavery in nineteenth century America has not saved us from becoming slaves to a massive system of production and consumption. We are already owned and operated by this massive apparatus as it guides our collective human destiny. The new abolitionist paradise engineers would have us believe that technology will bring an end to all suffering and pain. They are hoping that you buy the ultimate lie to ensure the complete enslavement of the entire human race. The freedom of the mind is the final frontier to be completely conquered by this system in order to achieve complete dominance over the planet.

-Chapter Six-
Eugenics

So where did this term, transhumanism, actually come from? Present transhumanists like to point to classic pieces of literature such as Dante's *Divine Comedy*.[39] This is quite accurate and it is true that the Italian verb transhumanar came from this source. In Dante's classic, the verb transhumanar is used to describe the act of going beyond the human condition, to transcend what it means to be human. This concept is no doubt a part of the philosophic bedrock of transhumanism, but to leave the explanation of transhumanism at this would be incomplete. We are talking about a philosophic, scientific, and religious singularity here. The fact of the matter is that the actual word transhumanism popped up in its modern scientific context in the middle of the twentieth century.

The man who first coined the term transhumanism and used it in its modern sense was Sir Julian Huxley. He was a well-known British naturalist and eugenicist who also happened to be the very first director-general of the United Nations Educational, Scientific, and Cultural Organization (UNESCO). He was a leading figure in the eugenics movement of the twentieth century. He was both president and vice president of the British Eugenics Society and a frequent award winner and speaker within eugenics circles. It can clearly be seen that his role as an internationalist combines with his work in eugenics. All one must do to realize this is read the preparatory document from UNESCO's 1946 inception, *UNESCO, Its Purpose and Its Philosophy* penned by Huxley himself. The ideological mission of eugenics had everything to do with a worldwide evolution of culture, biology, and government. This may be hard to follow, so here is a basic breakdown: the idea is to initiate a collective global evolution of mankind, to see that it takes a new and better form. Mankind is a biological, intellectual, and social being, so all of these aspects will have to evolve together. This process was and is a direct extension of humanism. A scientific, world humanism was eventually to be established by world revolutionaries so that a global evolution

---

[39] This reference certainly points to the tie between transhumanism and mysticism.

could occur. The ties between internationalism and eugenics are strong, with many world revolutionaries such as H.G. Wells being leading eugenicists.[40]

The ambitious eugenic plan was to improve the quality of the collective human genome. During the early twentieth century, the time-tested and revered practice of selective breeding was still believed to be the best method available of doing this. "intelligent love," was something that the eugenicists wished they could have everyone practice. The only way to get people to see the light of their eugenic creed was through education/indoctrination. The eugenic idea had to take hold in people's minds before any practical eugenics could actually be carried out. Huxley's preparatory document for UNESCO was, among other things, an outline for the worldwide spread of eugenic education. A standardized form of education would unify minds across the world and get them in line with the humanist agenda. This was all to be done in the hope of "discovering and pursuing the desirable direction of human evolution." The transhumanists would have you believe that Huxley's definition of transhumanism is somehow different from their own, but the more you actually read Huxley's material, the more you see remarkable similarities.

The following quote is from Julian Huxley's *New Bottles for New Wine*, the book in which he coined the term transhumanism within a segment by the same name,

> The human species can, if it wishes, transcend itself —not just sporadically, an individual here in one way, an individual there in another way, but in its entirety, as humanity. We need a name for this new belief. Perhaps transhumanism will serve: man remaining man, but transcending himself, by realizing new possibilities of and for his human nature.[41]

You should know that this part of Huxley's book was posted to the WTA's website in 2004 by James Hughes. Huxley's picture has also appeared directly on the homepage of the H+ site. H+ is merely the new name for the WTA; they are the exact same thing. As you can see, name changes happen frequently but ideas remain the same. Altering a brand name is a quick way to make people perceive change where there truly is none. Incomplete understanding of these topics actually makes them more marketable.

---

[40] Wells was also the head of British Intelligence during WWII.

[41] http://transhumanism.org/index.php/WTA/more/huxley

Eugenics

To understand what Huxley meant by *transhumanism* and what the transhumanist movement of today really is, one must understand eugenics. At its core, eugenics was (and is) about the self-guided evolution of mankind. It was not all vicious forced sterilization programs. It actually had the sincere hope of bettering the human race. This is incredibly important to realize, because it points to the very real danger of social/ political movements that call for improving anything. There is usually a larger reality that exists behind the curtain of public perception when it comes to these movements. To be effective any large movement needs a serious P.R. department in charge of keeping its image clean. The eugenics movement presented itself to the public as a glowing beacon of progress with exhibitions at state fairs and other public events in the 1920s. In their innocence, many people took these shows at face value and believed in their scientific merit. While "fitter families" events caught the attention of the public eye, there was serious work being conducted in laboratories by leading geneticists such as Charles Davenport. The Cold Springs Harbor laboratory in New York was established with big money from the Carnegie Institute and donations from Mrs. E.H. Harriman.

At the Eugenics Records Office in Cold Springs Harbor, huge amounts of data were collected and poured over for genetic research. The information gathered from questionnaires given out to average American families was used in conjunction with family histories. This was done to study specific genetic traits and how they were inherited within certain families. Charles Davenport understood that there were gigantic resources of information available to him already.

> While the acquisition of new data is desirable, much can be done by studying the extent records of institutions. The amount of such data is enormous,[42]

Records from hospitals, prisons, insurance companies, college gyms, and more were listed among these valuable sources of information.

> These records should be studied, their hereditary data sifted out and properly recorded on cards and the cards sent to a central bureau for study in order that data should be placed in their proper relations in the great strains of human protoplasm that are coursing through the country,

If Davenport had access to Artificial Intelligence do you think he would have used it? The remarkable thing to realize is that among the first major uses of AI is that of bioinformatics. From its inception, AI

---

[42] Davenport, *Eugenics: The Science of Human Improvement by Better Breeding*, 1910

is being used to pour over genetic data and sift out meaningful patterns.

We have to understand the kinder, gentler side of eugenics. Know that it was an idealist movement with good intentions. Its cruel method was rationalized. Eugenics' ends justified its means. Davenport sincerely believed that drafting legislation to prevent "unfit" persons from breeding would bring about a better world. The Eugenics Records Office lobbied for such causes. This marked a synthesis of science, politics, and ethics.

Davenport had this to say about the charity of eugenics,

> Vastly more effective than ten million dollars to 'charity' would be ten millions to eugenics. He who, by such a gift, should redeem mankind from vice, imbecility and suffering would be the world's wisest philanthropist.

It was well known by Davenport where the money for his work was coming from. He knew that powerful investors were on his side and this comment could be taken as a congratulatory nod to them.

The eugenics movement had its strongest support from extremely wealthy heads of industry and it was through major foundation money that its influence spread overseas. The Kaiser Wilhelm Institute for Psychiatry in Germany was funded in part by the Rockefeller Foundation. It was Rockefeller Foundation money that fueled much of the "psychiatric research" done by the Nazis. This is a plain fact that reveals much about the eugenics movement and so much more. Some of the most powerful people in the world sought a way to eliminate the bulk of the population that now caused them problems. Pesky labor unions had to be controlled with military force in those days. Individual human rights were a major drag on industrial progress. Workers who demanded higher standards hindered efficiency.

Through Social Darwinism[43] eugenics was legitimized. The supposedly natural hierarchical structure of human power was given scientific justification. Of course the wealthiest, most powerful people in the world sponsored this entire movement. They created it as an ode to their own dominance. Observing their power, they were already convinced of their superiority, but they needed a way to ensure that their influence did not diminish. They needed to be able to direct society in a way that was most beneficial to them personally. Justifying themselves as wise rulers was the perfect way to make their

---

[43] the application of evolutionary theory and principles to human society.

reign seem completely natural. Characters such as Charles Davenport were recruited to further their push to study the genetic traits of an inferior and servile population whose job it was to sustain the regenerate ones at the top.

The wise regeneration of human beings through controlled breeding is of course at the core of eugenic thought. The practice of positive eugenics ties in directly with the practice of inbreeding. Royal families are famous for inbreeding and eugenicists are keen to point this out.

In their collaborative book *The Science of Life* (hereafter called The SoL) H.G. Wells, son G.P. Wells, and Julian Huxley point out royal breeding practices, and they also mention other distinguished families that followed a similar reproductive regimen. One such example is the obsessive intergenerational breeding that occurred between the Galton, Darwin, and Wedgwood families. It should be noted that it was from these families that the base structure of our natural sciences was created, and the eugenics movement established! Both the theory of evolution and eugenics came from the same origin. This fact is something that gets completely overlooked by most people. The connection is conveniently not made in textbook history lessons. It was Francis Galton who first created eugenics in 1883 and of course, it was his cousin Charles Darwin who formulated the theory of evolution. Natural science as it stands today is a direct product of an inter-breeding group of families. The approval of this science came directly from the Royal Society with help from Julian Huxley's grandfather, Thomas Henry Huxley. This is a family affair if ever there was one.

The SoL says that inbreeding is actually a good thing. Because of predictability and decrease in variation of type, it is a wonderful method of reproducing genius.

> Charles Darwin himself married his cousin; and among his children and his grandchildren alike there have been several distinguished bearers of the same penetrating yet patient scientific capacity which characterized him, his cousin Francis Galton, and his grandfather, Erasmus Darwin,
>
> *-The Science of Life*, 1934

To the obsessive eugenicists not only was inbreeding acceptable, it was actually desirable due to its predictability. It was a mystic method of regenerating the self. Such human breeding techniques are similar to those used to produce dogs for specific purposes. Wells and Huxley also point this out in The SoL, a book which itself goes on about natural science for nearly fifteen hundred pages to end on a hardcore eugenic rallying cry in its final chapters. The point is made clear that the more "civilized" countries are adept at dog breeding. Wise breeders are quick to dispose of the inferior types that pop up during the eugenic process of perfecting the type. The destruction of the weak leads to a good breed overall. This is the eugenic justification for death, as well as for inbreeding.

They actually envision a eugenic "utopia" in The SoL (the soul?),

> Perhaps in years to come our descendants will look with intelligence over their pedigree, and if there is a probability of recessive genius in a family and no reason to suspect a grave recessive taint, they will deliberately encourage inbreeding. A rather grim Utopia might be devised in which for some generations, on the pattern of East and Jones' maize,[44] inbreeding would be made compulsory, with a prompt resort to the lethal chamber for any undesirable results. A grim Utopia, no doubt, but in that manner our race might be purged of its evil recessives forever.[45]

The dirty work of killing off inferior types is all part of doing a greater eugenic good. The fact that many people are ignorant slaves is explained in The SoL. Since they are a burden both to themselves and to society the most humane thing to do is eliminate them,

> People are indeed enslaved, sweated, fed on broken crusts, but they drag on, and they can procreate. Humanity is certainly accumulating a substratum of these dull unkilled.[46]

You see, slaves were actually useful in the old world, and they were fed just enough "broken crusts" to continue serving their masters. In the New World, an enhanced world that has gone beyond scarcity, these apparently subhuman types would no longer be needed. They would only get in the way of further progress.

> The world is passing into a new self-conscious phase of economic and social organization, which has little use for acquiescent drudges, and may develop an active impatience with merely consuming parasites and commensals.[47]

It is interesting to put this comment into the context of the technoprogressive transhumanist utopia that envisions a world that has effectively advanced "beyond scarcity." If everything becomes

---

[44] Experiments done on corn.

[45] Wells/Wells/Huxley, *Science of Life*, page 503

[46] Wells/Wells/Huxley, *Science of Life*, page 1409

[47] Wells/Wells/Huxley, *Science of Life*, page 1410

Eugenics

easily accessible and instantly provided for, then who wouldn't be considered a "merely consuming parasite?" We hear so much about how poverty is going to be eliminated, but how will this come about? What does it even mean to eliminate poverty? Who will own the technologies that bring abundance? These are important questions, as there is a definite conflict that has been pointed out time and again between massive population and extreme abundance. That conflict being that abundance cannot be given to everyone. That is to say, not without major conditions.

The many "defectives" throughout the population are viewed as a major hindrance to the progressive eugenic plan. By simply existing, these people slow down the entire process of forward movement. Any real advance in science and medicine would make it easier for all of these people to breed faster, live longer, and succeed. This certainly is not part of the eugenic plan. Instead, the idea is to cut these types off at the pass, to convince them to self-sterilize so that their numbers will drop. This will insure that the benefits of scientific progress go to those who are worthy of them. Education and propaganda could fulfill this mission, but even these powerful tools aren't always enough to finish the job. There remain a few types that just never learn. These "unteachables" as The SoL calls them, presented a major problem to the eugenicists of the time, so a plan was devised specifically for "birth control work in the slums." This is how it went,

> These unteachables constitute pockets of evil germ-plasm responsible for a large amount of vice, disease, defect, and pauperism. But the problem of their elimination is a very subtle one, and there must be no suspicion of harshness or brutality in its solution. Many of these low types might be bribed or otherwise persuaded to accept voluntary sterilization.[48]

And there is this famous quote by major eugenicist, and founder of Planned Parenthood, Margaret Sanger,

> The minister's work is also important and also he should be trained, perhaps by the federation as to our ideal and the goal that we hope to reach. We do not want word to go out that we want to exterminate the Negro population, and the minister is the man who can straighten out that idea if it ever occurs to any of their more rebellious members.[49]

It was assumed that the unteachables could eventually be bribed into sterilizing themselves. The success of this plan hinged upon its concealment. This is why today we see all kinds of eugenic bribes

---

[48] Wells/Wells/Huxley, *Science of Life*, page 1470

[49] Dec. 19, 1939 letter to Dr. Clarence Gamble

taking the form of social "services." A massive state-run system has been built up all around us and we haven't even realized it. This system has had as its goal, total control over the family unit. Reduced fertility has remained a serious political issue, albeit a concealed one, for quite some time now. Make no mistake the eugenic plan has been successful to a large extent. A self-proclaimed benevolent and progressive government has doled out effective bribes. This facade has been maintained, as the eugenic plan to perfect all life lies in the hands of an increasingly powerful state.

In the 1960s the Eugenics Society in Britain wrote up reports that expressed the need for "crypto-eugenics" programs. This would entail the furthering of their great work under different names. It was an effort to hide eugenics from the public eye by simply renaming it. The word *eugenics* had to be avoided completely because it was now politically incorrect. This simply meant that it was no longer polite to use the word *eugenics*, but bold action isn't about being polite. Only subservient weaklings are polite. This is why they bow to policy while the dominant types get to formulate it in the first place.[50] A sick power structure is at work here and its interests are protected through cunning and guile. Stealth is the defining quality of crypto-eugenics. Politics remains nothing more than a means by which a polite (obedient) society is maintained. Politics, policy, and a police force,[51] all this equates to a population convinced that its own slavery is actually a good thing.

In 1974, then Secretary of State Henry Kissinger wrote *National Security Study Memorandum 200*. This report, which remained classified for fifteen years, focused specifically on formulating world population control policies. Reducing the world's population was seen as a necessary action to secure international development, and American security,

> We cannot wait for overall modernization and development to produce lower fertility rates naturally,
>
> -Henry Kissinger, *NSSM 200*, 1974

A plethora of international NGOs along with the many United Nations projects focused on "population activities" were mentioned.

---

[50] Read Charles Galton Darwin's *The Next Million Years*.

[51] All have the same root: *poli*. All are about the creation and enforcement of policy (not law). Political correctness is simply adhering to policy without making a fuss; It is being *polite*.

Eugenics

A staggering amount of money had been put into these organizations but more would have to come. Results in fertility reduction would require that actual policies be made effective immediately. The typical Malthusian calamity[52] was the impetus to actually wage a silent war on the world's population. The very necessities of life, food and water, could be used as a weapon to see this eugenic mission through,

> A growing number of experts believe that the population situation is already more serious and less amenable to solution through voluntary measures than is generally accepted... even stronger measures are required and some fundamental, very difficult moral issues need to be addressed... our own consumption patterns, mandatory programs, tight control of our food resources explicit consideration of them should begin in the executive branch, the congress and the U.N. soon,[53]

All sorts of tricks to get people to reproduce less were mentioned. Social programs of all kinds are effective and the fact is that economic development itself reduces fertility. NSSM 200 said that increased education could lead to higher employment, and corresponding reduced fertility. Because of this, social programs involved in the "broadening of nontraditional roles for women" were advocated. Any situation that makes it difficult for humans to reproduce will help the eugenic cause of population reduction. It seems strange to most people that government programs designed to "aid" families are often designed to lower fertility and create a dependent population. Again, NSSM 200 said,

> Because the family is the basic unit of society, governments should assist families as far as possible through legislation and services.

This equates to the bribery mentioned by Wells and Huxley. The crypto-eugenic plan and its success rest completely on its effective concealment. Clever language is deployed to convince the targeted public that its interests are actually being served. The full implication of these various programs remains concealed. The more families come to depend on social services, the less they will have to depend on each other. The destruction of the family unit as we know it is all part of this worldwide project. Kissinger wrote,

> World population growth is widely recognized within the government as a current danger of the highest magnitude calling for urgent measures,[54]

---

[52] Thomas Malthus, a pioneering voice for population reduction.

[53] National Security Council, *NSSM 200*, page 14

[54] National Security Council, *NSSM 200*

Little did the public realize at the time Kissinger wrote this that by simply reproducing, they were considered a problem.

Julian Huxley knew what he was doing. He knew that UNESCO could achieve its eugenic goals if given enough time and money. He clearly explained this as he said,

> it will be important for UNESCO to see that the eugenic problem is examined with the greatest care, and that the public mind is informed of the issues at stake so that much that now is unthinkable may at least become thinkable.[55]

Human tradition has held eugenics back. Family life as we know it is eroding, and as it does, we remain unaware of the reasons why. We haven't been told about the various agendas formulated and implemented at an international level. We never learned about the eugenics movement in American History class. We don't know that words such as "family planning" are nothing more than creations of the eugenicists' PR people. It never occurred to us that our lives are viewed as commodities. The value of our existence now lies beyond the bounds of our very humanity. The final goal of world revolution is to transhumanize the population. To do away with the inefficient, stupid, and dirty unwashed masses in favor of genetically engineered posthumans. Absolutely no love for humanity is held by the eugenicists, or by many leading transhumanists, for that matter. Reading transhumanist literature, you will find that they frequently point out how weak and pathetic humans are. The supposed scientific fact of the matter is that we have reached a natural evolutionary plateau, and we must now use technology to speed up the process.

The absolute fact that an intellectual elite directs the world is admitted to in The SoL,

> The inventions and organizations that have produced the peculiar opportunities and dangers of the modern world have been the work so far of a few hundred thousand exceptionally clever and enterprising people. The rest of mankind has just been carried along by them, and has remained practically what it was a thousand years ago. Upon an understanding and competent minority, which may not exceed a million or so in all the world, depends the whole progress and stability of the collective human enterprise at the present time. They are in perpetual conflict with hampering traditions and the obduracy of nature. They are themselves encumbered by the imperfections of their own trainings and the lack of organized solidarity. By wresting education more or less completely from its present function of transmitting tradition, they may

---

[55] Huxley, *UNESCO: Its Purpose*, page 21

be able to bring a few score or a few hundred millions into active co-operation with their efforts. Their task will still be a gigantic one.[56]

This is right in line with the *Open Conspiracy* that H.G. Wells was so enthused about. He wrote an entire book dedicated to the subject. Worldwide education is the method by which the eugenic idea is to become acceptable. The recruitment of "Open Conspirators" is incredibly important. The transhumanist movement today is absolutely an aspect of this Open Conspiracy. It has found willing and able minions, mainly from academia, to promote the progressive idea of speedy self-directed evolution.

The subject of a dominant minority was approached directly by Mike Treder, who is a leading transhumanist himself. In his article *The Power Pyramid*, posted to the IEET website in 2008, he probed interesting territory. He made a pyramid graphic to illustrate his contention that out of six billion people in the world there is an elite group of only 6,000 individuals who hold true power. This makes them literally one in a million. Below them are the "intellectuals, academics, and activists" who are one in a thousand. Out of this pool of motivated idealists come the Open Conspirators that H.G. Wells so vividly portrayed as the spokesmen for world revolution. Treder says the very same thing in this article. There are only six million people in the world who are up to such a task. What this group can effectively get done rests in the hands of an even smaller super-rich segment of the population that holds controlling interest over society. A dilemma arises for Treder as he realistically questions what people in his position can do to effect change in the world without completely succumbing to the will of the Power Pyramid's capstone,

> So, what does this mean? How can those of us in the small green segment of the global pyramid above make an impact on the much tinier red segment at the top? Is there any real hope for that? Or are we just spitting into the wind as we hold forth with our opinions?[57]

The eugenic problem of the transhumanist movement is not completely cut and dry. This is why so many individuals cannot see it. The truth is that eugenics is a big part of the transhuman project. Eugenics involves destruction of the weak, strict population control policies, and improvement of the collective human gene pool. It is an elitist ideology through and through, a completely ruthless plan that cannot be openly advocated. If the whole truth were broadcast to the masses it would alert them to their previously unknown enemy. The

---

[56] Wells/Wells/Huxley, *Science of Life*, page 1471

[57] IEET, Treder, *Power Pyramid*, http://ieet.org/index.php/IEET/more/2469

only way to promote eugenics effectively is to find the right Open Conspirators who are motivated by their own particular ideals. Basically, a group of highly educated lackeys must be amassed to influence the rest of the population through proclaimed education. It seems as though Treder picked up on this in his article, and lamented the true scope of his own predicament.

The SoL explains how the intellectual type could be used to insure the success of eugenic progress,

> For a number of generations, at any rate, a dead-weight of the dull, silly, under-developed, weak and aimless will have to be carried by the guiding wills and intelligences of mankind... obsolescent religious forms and plausible political catchwords will be used to rally and canalize their mental weaknesses. The brighter, more energetic types of stupidity and egotism will be constantly organizing and exploiting the impulses and uneasiness of this universally diffused multitude for here we are not writing of any social class or stratum in particular. The inferior sort is found in greater or less abundance at every level.[58]

Bringing About Collective Evolution

Eugenics is regarded today as pseudoscience, but there was a time in which it was considered legitimate and accepted throughout academia. Books such as *The Science of Life* were published specifically for a peer group of learned men. Eugenics took hold among this crowd because it catered to their egos. These professors could count themselves among the intelligent few who were encouraged to reproduce. Of course not everyone was on board with eugenics, but only a small minority of individuals were to be found within academic circles that opposed it. It was almost entirely accepted as fact and used as a basis for further scientific work in genetics. Textbooks called for eugenic policies. Social programs were designed with genetic science as their justification. Social Darwinism took root and it did not happen by chance. There were very powerful people behind the eugenics movement. For them, eugenics was part of a cosmic philosophy. It continues to be so to this very day.

Eugenics was a way of joining the natural and social sciences. Eugenics, taken to its desired ends, would end up in a technocracy. A technocracy is a government that is led by enlightened scientists and bureaucrats. A technocracy is an ideal state that bases all political decisions on applied logic and reason. In theory it would be

---

[58] Wells/Wells/Huxley, *Science of Life*, page 1471

Eugenics

completely fair and objective in its authority. The justification for such a system lies in the wisdom of science itself, and in this way a technocracy stands at the mercy of any bad science that it may wield. A famous fictional technocracy is seen in Aldous Huxley's *A Brave New World*. It was his brother, Julian, who literally wanted to use the scientific method to "increase in our knowledge of and control over the phenomena of human and social life." But serious changes would have to come about if any real progress was to be made along these lines,

> The dead weight of genetic stupidity, physical weakness, mental instability, and disease proneness, which already exist in the human species, will prove too great a burden for real progress to be achieved,
>
> -Julian Huxley, *UNESCO: Its Purpose and Its Philosophy*, 1947

A technocracy seeks to oversee every aspect of human life. Maintaining a level and efficient population is an important aspect of any serious technocracy. UNESCO had as one of its goals the eventual establishment of this sort of population control,

> The recognition of the idea of an optimum population-size... is an indispensable first step towards that planned control of populations which is necessary if man's blind reproductive urges are not to wreck his ideals and his plans for material and spiritual betterment.[59]

In these quotes from *UNESCO, Its Purpose and Its Philosophy* you can see ties to current transhumanism and genetic science, but this next quote is particularly telling,

> From the evolutionary point of view, the destiny of man may be summed up very simply: it is to realise the maximum progress in the minimum time.[60]

Now, what is transhumanism if it isn't the self-directed and speedy evolution of man? I would say that this phrase defines transhumanism to a T. Huxley's idea about the destiny of Man is nearly identical to the following, which is the very first sentence of *The Transhumanist FAQ* document,

> Transhumanism is a way of thinking about the future that is based on the premise that the human species in its current form does not represent the end of our development but rather a comparatively early phase.

---

[59] Huxley, *UNESCO: Its Purpose*, page 45

[60] Huxley, *UNESCO: Its Purpose*, page 12

The need to enhance is born of the desire to speed up our own evolution so that we may quickly become a higher form of life. Transhumanists, however, are quick to refute their association with eugenics. Prominent transhumanist Natasha Vita More has claimed that Huxley's version of transhumanism has to do with improving humanity while still remaining fundamentally human. It is her argument that the present concept of transhumanism has to do with going beyond human limits through the use of technology.[61] This specific point supposedly diverges from Huxley's own ideas about transcendence, and eugenics as well. Eugenics was purely genetic whereas transhumanism is much more. This is a very fragile argument at best, and one has to wonder about the new H+ name that has been given to transhumanism. It is in this name change that we see an even closer resemblance to Huxley's ideas, not a divergence. All the marketing tricks and denials in the world can't hide the truth. Top transhumanists are smart, so they know that transhumanism is indeed a direct descendent of eugenics. This is why they give out awards named after H.G. Wells and J.B.S. Haldane. These men were influential revolutionaries and eugenicists. The reason that the truth about eugenics cannot be admitted to overtly is because of the eugenics movement's political baggage. Any mention of eugenics is now bad PR, but the idea of a world (r)evolutionary movement is now more prevalent than ever. That is why global governance and progressive political concepts are discussed all day by transhumanists. They are taking elements from the works of H.G. Wells that are still politically correct and using them to promote a

---

[61] At one time the specific article mentioned here was online, but when searching for it to list in the bibliography of this book it was not to be found.

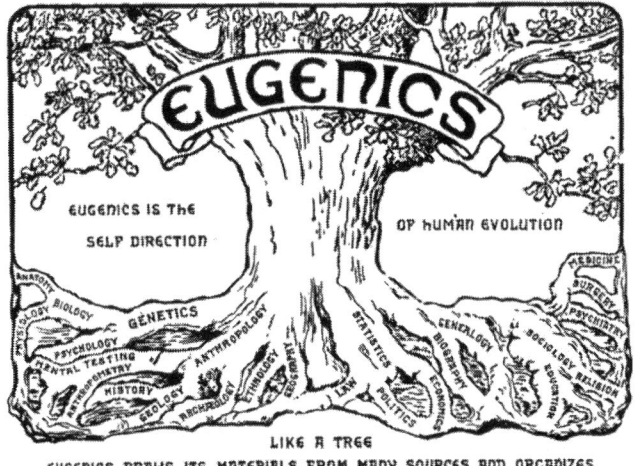

Fig. 6.1 Illustration from *The Second International Congress of Eugenics*

better future. Eugenics never actually went away; it simply shape-shifted into new, more progressive sounding movements.

In terms of genetic engineering, and the further use of science to bring about rapid and improved evolution, the eugenicists talked about such things long ago. The SoL had this to say about the future of eugenics,

> At present eugenics is merely the word for what still remains an impracticable idea. But it is clear that what man can do with wheat and maize may be done with every living species in the world - including his own. Because at present our knowledge of genetics is too limited to do more than define certain sorts of union as 'undesirable' and others as 'propitious,' it does not follow that we shall always be as helpless. There may come a time when the species will have a definite reproductive policy, and will be working directly for the emergence and selection of certain recessives and the elimination of this or that dominant. In our treatment of genetics we have given a few first-fruits of the science, which suggest what forms the practical eugenic work of the future is likely to take. Once the eugenic phase is reached, humanity may increase very rapidly in skill, mental power and general vigour.[62]

---

[62] Wells/Wells/Huxley, *Science of Life*, page 1478

Nobel Laureate Joshua Lederberg spoke to the subject of eugenics, as it related to the growing science of euthenics. Euthenics involves improving the genetics of a living creature while it is still developing. Before being born it may be genetically modified for improvement. Lederberg said specifically that euthenics would increase the speed and efficiency of eugenics. He also mentioned the fact that it would bring about a faster rate of evolution.

It's been said that a picture is worth a thousand words. In the case of this illustration (Fig 6.1), I would have to extend that number a great deal further. A whole book could be written on this Eugenics tree. For one thing, the ancient quest for knowledge is symbolized by the tree symbol. This particular tree most definitely depicts that quest and puts it in the context of eugenics. Eugenics is seen as the trunk of the tree, while many divisions (aka branches) of human endeavor make up its roots.

> Like a tree Eugenics draws its materials from many sources and organizes them into an harmonious entity. (Fig. 6.1)

The point is reiterated that many seemingly separate establishments culminate into a singularity. It is through unity that these pools of knowledge can grow into a powerful creative force. The goal of both the transhumanist and eugenics movements is convergence. The tree tells us that "Eugenics is the self direction of human evolution." Does this sound familiar?

The transhuman love affair with global governance ties in directly with the eugenics movement. It is the singular "entity" (Fig. 6.1) of a world state that is being created by this massive operation. The world state is to be run scientifically. The knowledge of human behavior and biology will be used to establish an effective technocracy. Notice that this eugenics tree has politics, law, and surgery among its roots alongside other favorites such as mental testing, education, and psychology. Exactly how do you think these disciplines would ultimately combine into one "harmonious entity?" When one really thinks about this tree, the vision of a dystopia is hard to avoid. But one man's dystopia is another's utopia. It all depends on where one lies on the scale of social status. Under such a socialist system, the best and brightest would be the owners/ operators of the show, and they would be sure to arrange all other useful types in a logical descending order of importance based on IQ tests (among other things). Positions of influence would be doled out logically.

Converging technology works in the same way as the eugenics tree. Many different sciences combine to form a singularity. The idea

Eugenics

of singularity has been around for a long time and has been reintroduced to illustrate a powerful idea. Transhumanism is the new, high tech, flashy way to present this concept to people. With the advent of new technologies, this mission continues to be easier to carry out. It is very telling that IBM produced punch card computer systems for the Nazis. International Business Machines were used to record eugenic operations and to make a horrible administration more efficient. Eugenics is an International Business that has always been run with the help of Machines. Today, artificial intelligence machines are allowing the interpretation of huge amounts of biological data.

The largest AI projects have been directed toward understanding human genetics. Major government departments have funded these. "Post-Genomic bioinformatics" is a huge project that builds off the work of the Human Genome Project. The transhumanists are here to put a good spin on this massive undertaking. They want us to understand that improving ourselves through genetic engineering can only be a good thing. The ability to direct the biology of our offspring will lead to a better quality of life. We are being led to believe that enhancement is an end in itself. The consequences of this enhancement are known but not seriously addressed by transhumanists. It is their job to sell the idea, not to make it work. Life extension is a hard thing to oppose and so it is incredibly easy to elbow your way to the moral high ground by simply advocating it. Life will be both genetically altered and extended in duration, but to balance the scales of reason, Huxley's "idea of an optimum population-size" will also have to be sold. The full magnitude of population reduction is not something that transhumanists address. We aren't ready to hear about that yet. For now, we need to be sold on the benefits of technological enhancement.

At a point in which life extension, cognitive upgrades, and other technologies are made publicly available, what exactly will be the condition upon which they are given out? Seriously, we can't expect a blank check for these things. A price undoubtedly will be paid for such monumental changes. When human life can be extended indefinitely, what will happen to the size of the population? We have already been bombarded with propaganda telling us there are too many of us. What would happen if people began extending their lives while continuing to reproduce at the current rate? It has been said that life extension would make it possible for us to stay physically young indefinitely, so reproduction could continue even upon old age. This is an obvious problem that has a fairly obvious answer, but it is not an answer that anyone wants to hear. The fact that human reproduction will have to be controlled or even replaced by a better scientific alternative is not often discussed by transhumanists,[63] because it scares people off. It is also another hint that transhumanism and eugenics are one in the same thing. In the short term transhumanism has to convince people that life extension is a generally good idea. When actual life extension therapies arrive on the scene the precedent will be set for governments to come in and manage the population size. Now how will this jive with the natural human freedom to reproduce independently? In the absence of reproductive freedom, will we have to be content with the liberty to die when and how we want?

The life and death dialectic would ultimately synthesize by introducing one-child policies or similar technocratic laws. This of course would first be debated on television to get our minds familiar with the idea, and then followed quickly by the democratic process of accepting actual policies as they are written into law. This is one small portion of the price to be paid for life extension.

The definition of freedom in a transhumanized utopia is a difficult thing to envision. Would one be free to opt out of the various upgrades that are made available, and what would be the ramifications of doing so? In a world filled with enhanced minds, how would anyone be able to function without enhancing? What kind of Social Darwinian competition will manifest by simply enhancing in the first place? More than this, what will it actually mean to enhance? These problems have already made themselves apparent in our modern world. The difficulty of keeping up with technology can be a very serious psychological burden, especially for aging people. This trend would be taken to its extremes if we choose to go all the way with enhancement. What exactly are we getting from technological progress? Is the luxury of captivity really better than the freedom of the wild? The transhumanists like to gloss over any and all of these concerns. They want us to believe that progress is an inherently good thing. There is supposedly no other option but to go forward. It is inevitable that the human race do these things because they represent our natural and collective destiny. This natural destiny entails using the art of science (alchemy) and abandoning the slow evolution of nature itself.

A system that provides everything is one that fosters dependence. Independence is marked by the ability of the individual to do things for himself. The eugenic plan is to create a state so vast that it has the

---

[63] You have to read their white papers to get that story.

Eugenics

power to control every facet of life. In this way it would gain total control over the individual. This is why the sales pitch for this system has to be presented as democratic. The illusion that some sort of collective good is being done is effective in convincing people that everything is OK. A process of destroy and rebuild is now in effect. The eugenicists spoke to this powerful truth as well.

Quoting again from *New Bottles for New Wine* Huxley said,

> This process too will begin by being unpleasant, and end by being beneficent. It will begin by destroying the ideas and the institutions that stand in the way of our realizing our possibilities (or even deny that the possibilities are there to be realized), and will go on by at least making a start with the actual construction of true human destiny.[64]

Old institutions and ideas have to be destroyed to see this plan through. This is not an easy task and men like Huxley are well aware of what they are up against. People are wary of change, and with good reason. Mass acceptance of eugenics is the endgame of the elite. Do not underestimate the lengths to which this group will go in order that their will be done. They know how to effect change and obtain influence one way or another. In an article written about Julian Huxley that appeared in the Galton Institute's December, 1999 newsletter, the writer pointed to the fact that ideas themselves take time to evolve. Quite chillingly he suggested the following,

> A catastrophic event may be needed for evolution to move at an accelerated pace, as the extinction of the dinosaurs gave the mammals their chance to take over the world. It is much the same with ideas whose time has not yet come; they must survive periods when they are not generally welcome.

The old must die so that the new may live. It is through a massive act of violence that evolution would supposedly speed up. The correlation between catastrophic events and evolving ideas is extremely disturbing to say the least. This actually characterizes the alchemical method that has been employed upon the mass mind itself. Through alchemy our minds have been changed, and where the mind goes, the body follows. The ultimate goal of eugenic evolution is that of complete biological control.

Leading transhumanist and AGI researcher, Ben Goertzel, has said that "we've evolved to be relatively stupid creatures." It is his contention that humans would be wise to allow a powerful AGI with a "more rational mind" to take control over our world. Such an AGI wouldn't be as "insanely emotional," and wouldn't have the problems

---

[64] http://www.transumanism.org/index.php/WTA/more/huxley

associated with our "chimp-like motivational systems." From this point of view, no human is viewed as particularly intelligent or useful. Theoretically, once produced, an AGI may have a similar mindset and decide to carry out methodical genetic experiments in order to improve human beings in a logical manner. This would be the cold, calculated, *intelligent love* of the perfect eugenicist. Speculation aside, a literal machine mind completely in tune with reason, and uninhibited by trivial emotions is the perfect metaphor for the Eugenicist himself.

There is a very important mystic quality to eugenics. The modern idea of eugenics has merely arisen from elitist philosophies.

> Because men that are free, well-born, well-bred, and conversant in honest companies, have naturally an instinct and spur that prompteth them unto virtuous actions, and withdraws them from vice, which is called honour.   -François Rabelais, *Gargantua*, 1534

Rabelais was an influential French occultist and hero of Aleister Crowley. It is from Rabelais that the occult philosophy of Thelema was created. It had everything to do with the power of the Will. This philosophy influenced the German Brotherhood of Saturn in 1928. They advocated the eugenic concept of "compassionless love."[65] Opting to obey intelligence instead of weakness and vice, these men surely chose to breed well.[66] All emotion must be destroyed so that divine will may be made manifest within the material sphere. Eugenics is the scientifically guided regeneration of men who are obsessed with their own mystic quest of becoming god. God is so intelligent and so powerful that *he* merely has to will something into existence. This is how he[67] procreates. Dark minds are envious of this

---

[65] Skeptics would argue that the word *compassion* doesn't fit in this context, but it does. Break it down to *com* as in *com*mon, *com*moner, a mixed and mingled person of *com*mon stock genetically, and *passion* actually meaning suffering. The other instance of passion is the love practiced by commoners, which is viewed as suffering esoterically. Having *compassion* one empathizes with the suffering of the common man, which is derived from his ignorance. *Love* itself is symbolic of regeneration. High esoteric love is for those who are above *Les Miserables*.

[66] The rites of the Brotherhood of Saturn speak to both "brothers" and "sisters" of the lodge. According to Stephen E. Flowers the brotherhood was involved in sexual occultism, and the order was "unabashedly Luciferian."

[67] It isn't right to refer to God as *he*, but in this context it fits.

power, and seek it for themselves. This is the mystic goal of eugenics. It involves going beyond the bounds of natural human regeneration.

Part Two
Digging Deeper

-Chapter Seven-
Logic and Reason

Scientific achievement and humanist ideals are said to be born out of logic and reason. Being an extension of humanism, transhumanism attempts to latch onto reason itself as a rationalization for its own particular points of view. In our time, reason has risen to a level of purity that rivals that of religious ecstasy. It has slowly risen to prominence as the holy spirit of a new faith. Science is the body of this faith. Be aware that no body can function without a head. Who are the brains behind this operation? What is the nature of logic itself?

The idea that science and technology should be used to improve human life is certainly a reasonable thing to say. Saying that it is only logical that human beings will merge with artificial intelligence machines so that they can live eternal lives and become godlike space travelers is another thing altogether. Making these sorts of claims in the name of reason is not very reasonable at all. They sound an awful lot like the heavenly promises of religion. Because of our learned, dualistic perspective on science and religion, it can be hard to make this association. We have to broaden our perspective and destroy any bias to see the true essence of logic. Pure logic and reason have no boundaries. In truth, there are an infinite number of directions that one can travel within these vehicles. The first logical step taken will affect the course of the entire journey. Where we are at today is the culmination of a specific set of directions, a certain logical map. We could just as easily have gone off in a completely different direction to exist in a reality very different from the one we find ourselves in now. Whatever that reality might have been does not matter; to anyone existing there it would have appeared to be the logical outcome of past experience. This would have been the case no matter how bizarre it would seem from our perspective. The ridiculous idea that progress inevitably leads forward, that natural evolution is a one-way street, is used to substantiate everything around us. This is done to make us think that everything is normal, when in reality there is no such thing as normal. There are individuals in this world who understand this power of perception and use it to their advantage.

128   Revolve

Fig. 7.1 Here we see the Inner School of the Mysteries being guarded by men with giant pens. They symbolize the exoteric University system and its professors. Their words are broadcast to a wide audience. The real truth remains hidden in the dark, and accessed only by those who are deemed worthy.

Anyone with the authority to define normalcy has amazing creative power. Over time he can make up new norms according to the direction he plans on taking the world in the future. Ideas shape reality. This explains the meaning of mind over matter, which is the essence of alchemy.

There is of course a problem in defining logic. We always end up with a bias of some sort. We have to wonder, what exactly is logic? It appears to be whatever the smartest guy in the room says it is. This is an intellectual version of might makes right. When we don't question an expert's opinion, we succumb to the might of their particular brand of logic. What the professor tells us magically becomes fact. We must understand that a profession is not necessarily a fact, it is an opinion, but a professor is a person who is put on a pedestal within society. His opinions are important. It isn't considered reasonable to question

Logic and Reason

them. This puts us in a dangerous situation. As soon as we stop questioning, we begin repeating rhetoric. The complexity of rhetoric really doesn't matter. What does matter is whether or not we use critical thinking to come to our own conclusions. Education that promotes critical thinking certainly enriches our lives, but do institutions really encourage free thought or are they just information-downloading stations? Are students repeating rhetoric, or are they truly learning? Do we continue questioning throughout our lives or at some point do we begin to faithfully listen?

The subjective quality of logic leads us to a profoundly important question. Who is in charge? Who is leading the technological push forward, and why are they doing it? There are answers to these questions for anyone who is willing to sincerely seek them. One answer is that knowledge is power, for it is only through knowledge that one can create. Creating technologies to organize *agents* (a term used frequently in computer, cognitive, and social science) allows one to obtain greater power over those agents. There has to be a method for deploying such technologies that seems reasonable. It is surprisingly easy to package ideas as reasonable; this is called marketing. All you have to do is appeal to popular opinion. This is how politics work, and it is how transhuman technologies have to be promoted so they will eventually be accepted and used by average people.

There are definite political, monetary, and social changes being planned. There are reasons for all of these changes that reach far beyond the vague concepts of progress and change that fuel the fires of popular opinion. Appealing to the crowd is a must for transhumanists, and like all other PR campaigns, theirs doesn't have to be completely logical to be effective. They are propagating an idea as widely as possible. All details are secondary to the main goal which is sowing the seeds of change. Questioning emerging technologies is extremely difficult, but it must be done. To simply say that every logical person will want technological/ biological enhancement is certainly not reasonable. There are many legitimate reasons that one may give to argue against the speedy development of such technologies. These legitimate concerns are being glossed over and demonized.

The crucial point lying behind all of this is that logic and reason depend on subjective bias. They cannot be separated into completely objective facts; there will always be some subjective influence. When logic and reason legitimize themselves, intellectual authoritarianism has come.[68] We cannot take the definition of a word such as logic for granted. This is the profound danger of any mass movement. The individual is required to give up his own unique ideas so that his mind may fit into the structured reason of a collective. This can occur to varying degrees but the more cohesive the group, the more collectivized its ideas become. Think about any given *fact*. To exist, facts have to be built up by objective evidence. This sounds good in theory, but how does one completely overcome bias when establishing facts? The initial information used to gather new information is supposedly objective, but this is not always true. Personal preference directs the focus of the scientist. While this can occur either consciously or unconsciously, the fact remains that subjective tastes play a key role in the direction of scientific study. This rule applies to whatever the predominating ideas of the time happen to be. More than this, scientists are very hesitant to pursue any ideas that don't fit within the accepted rules of their day. This makes real progress difficult and it keeps the directive force of logic on track.

Not only is science subject to these flaws, mathematics is as well. Geometry, the most revered form of mathematics to Freemasons and mystics alike, reveals within itself the mystery of subjectivity. There is a very profound reason that geometry has been held in such a high regard by Mystery Men. The highest, most illumined initiates, the cult leaders such as Pythagoras, have understood that their own sacred geometry represents nothing more than one pattern in an endless sea of other possible patterns. Its shape depends on the will of its creator. To the profane and lower degrees of initiates, the identity of this creator is explained as God itself. This explains the capital G within the compass and square as being both God, and Geometry simultaneously. In this way the high priests have given their own mathematical formulas a divine quality that is undeserved. Geometry is not sacred. That is to say, there is not one superior form of geometry. For indoctrinated minds this fact defies belief but it is true. Conventional wisdom places people under the influence of specific patterns. These patterns influence minds to such a degree that the ability to even question them becomes a monumental task.

For centuries Euclid's axioms stood as the undisputed gospel truth of geometry. It is incredibly fitting that his series *Elements* was written in 13 volumes. The mystic connotations of this are staggering. The number 13 itself deserves an entire book to explain.[69] The title

---

[68] James Hughes has written about this in his *Problems With Transhumanism* series at IEET.org

Elements certainly refers to the four elements of nature, the four seasons, and the four chief points of the zodiac (all of which are one and the same). In his title, Euclid was proclaiming his theorems to be the mathematic explanation (or decoding) of Nature itself.

It is incredibly significant that the Pythagorean Theorem plays an integral role within Freemasonry. It is important within the initiation ritual of the Master Mason; it is how one "squares the lodge." A square, of course, has four sides. The perfected ashlar, aka the perfected man is squared on all sides. This is the perfection that Sacred Geometry bestows, but how sacred is it?

In the nineteenth century several men in different countries created their own forms of Non-Euclidean Geometry that actually contradicted Euclid's postulates. His fifth postulate would not work within the framework of Hyperbolic Geometry, which was a creation of Nikolai Ivanovich Lobachevsky, a Russian. At nearly the same time a Bulgarian, János Bolyai, published a treatise on Hyperbolic Geometry as well. Bernhard Riemann also created a self-consistent and valid form of Non-Euclidean Geometry in 1868. A lot of action was going on in the field of Non-Euclidean Geometry at this time. Whatever the reasons for such parallel discovery, if there are any, the fact remains that the long-held truth of Euclid was taken off its pedestal. Since the time of the Greeks, Euclidean Geometry was the only accepted form of geometry. This gave it an authority that profoundly impacted the minds of men for centuries. These minds in turn shaped the world according to their knowledge, or more appropriately, their specific logical formulas. The power inherent in Euclid's pattern became the creative power of the universe[70] for centuries. It was the mathematical base upon which the cathedrals were built. You must realize the profundity of this fact. It reveals the essence of the esoteric religion of the enlightened ones. This religion has spawned all other faiths within this world.

In ages past the truth[71] was always freely given out to the masses in the form of religion. Whatever brand of truth happened to rule at any given time was of little consequence, because whatever its particulars happened to be, they were always absorbed by the people. There were, of course, individuals who did not succumb to the

---

[69] Specific numbers all have their own mystic significance. Numerology is very important within esoteric religion.

[70] This was only true within the minds of men.

[71] Again, *truth* is the wrong word to use here, but that is the point.

religious dogma of their day completely, but regardless, religion still held enormous power. Religious rituals were always attended by the mass. The power of ideas animates nations. Matter can attain to many different *states* of equilibrium. The state itself relies on the standardization of ideas. Without this it could not exist, let alone grow. This is why church and state have been so closely related throughout history. They are simply two fragments of an overarching system, two sides of one scale. Religion brings about a hardening of ideas. It can take any form that is necessary for it to achieve a certain goal. It is extremely humbling to reflect on the number of people who have come and gone from this world. So many of them convinced themselves that their religion represented ultimate truth and that its most literal interpretation was fact. What on earth is the meaning of this? How can people be made to believe false truths over, and over.

In today's world we like to congratulate ourselves on our collective battle against ignorance. Ignorance, which now uses the scapegoat of religion, is all but conquered, or so we're told. Sure, there are still fundamentalist thinkers, but they are only made up of uneducated simpletons. If only we could bring education to the world, then science could clear up dogmatic thinking once and for all. This is an idea that is becoming increasingly popular among intellectual elitists at all levels of society. This has not happened by chance. Where we are at today is the product of centuries of building. The building of logic has led to new forms of reasoning. Such reasoning has been embraced throughout all social castes. You have the unwitting know-it-all intellectuals at the bottom who see themselves as *bright* beacons of light in a world full of *dim*wits. At the top you have the Master Alchemists who craft opinions for mass consumption through the use of their magic, i.e. academia, media, the arts and sciences. It isn't uncommon for profane intellectuals to advocate atheism, but of course this is only done effectively by first attaining a high level of self-righteous indignation for anything or anyone dumb enough to believe otherwise. Atheism just makes sense because there is no proof of the existence of god. The conviction required to make such a bold statement is nothing to take lightly. Atheists don't seem to have a problem putting absolute limits on the unknown. In this way, their ism is just another faith. God has been unfortunately entangled with religion and in the process, both have been ruined. However, the real truth remains. The existence of god cannot be verified in any absolute terms (i.e. yes or no). Things are a bit more complex than that. Humbly accepting our limited capacity to know truth doesn't offer much reward for the ego. That is why many people choose the ism path as their road to truth. You'll be hard pressed to explain the religious appeal of atheism to atheists because they honestly view their ism as the antithesis of religious dogma. If they were aware of the Hegelian dialectic, they would understand that both thesis and antithesis are used to create a new and more complete whole. Instead of questioning their own method of reasoning they would rather get off on being anti-religious. These are the new believers for a new kind of religion. The mystic sowers of the seed are cultivating new religions. These philosophic sowers are planting ideas within the mass consciousness so that they will spring forth as a bountiful harvest. Are you under the impression that atheism has sprung up from nowhere?

Logic and reason can be used to justify a particular set of ideas. In the case of transhumanism, they are used to advocate the technological enhancement of the human race. The fact that transhumanism is described as "a product the European Enlightenment" is extremely important because profound historic and symbolic significance lies behind this. Top transhumanists are pointing to the evolution of their own ideas when they mention this time period. They know that the rational thought that began with Greek philosophers has, over time, grown into their own transcendent philosophy. They stand on the shoulders of master philosophers, who knew how to appeal to human nature. Remember that the high esoteric sciences have always been concerned with the study of Nature, inner human nature being most important. This is a necessary understanding one must come to realize in order to become a Master Alchemist. The powerful human mind is what sets human nature above the rest of the dark natural world. Such high intelligence has led to probing questions about life. Humans have always wanted answers. The philosophic elect have never ceased to exploit this innate yearning for truth. They have built up complex patterns of thought that supposedly lead to truth. Remember that Pythagoras was not only a craftsman of geometry but also of philosophy. He understood that thought could be directed with mathematic precision. His cult and others that followed laid the groundwork for the time period known as the Enlightenment.

It is important to understand that time periods can be symbols. The very name *Enlightenment* has extremely important symbolic meaning. Within the mystery tradition the idea of enlightenment is ubiquitous. Enlightenment is the goal of the initiate. It is the divine principle that is first contemplated and later emulated. To really understand the importance of what happened during the European Enlightenment, one must first understand the occult meaning of the word enlightenment. En (within) Light (cosmic radiation, divine energy, life itself) En (within) Ment (Mind, mental.) The word refers to an innate and divine creative energy that exists within the mind. Mind refers to both the macrocosmic mind of god, and the microcosmic mind of the individual man. This process is the way for an individual to make contact with the divine mind of creation. Attaining Enlightenment involves going to the darkest depths of your own mind to ultimately find the light within. All of the ancient creation myths speak of a dark chaos that existed before anything else. The coming of light marks the beginning of creation, and of life. Light, love, and logos are all terms that describe the creative force of god. The creator god uses light as his seed.

The actual physical manifestation of light comes from our star, the sun. Its radiation is what brings life to this planet. Our entire physical existence depends on light. Because of this, the sun itself has always been a profoundly important symbol and object of worship. It is revered within the Mysteries as the physical representative of an unseen creator. The sun has been idolized all throughout human history (and likely prehistory). Mythical heroes, prosaic representations of the sun, have taken on lives of their own within the minds of men. They have all been known as the Son of God, which is the same thing as the Sun of God. For example, Jesus Christ was a human symbol of the sun itself. He was the Sun of God/ Son of God. He is not the only hero to fit this template. It has always been these enlightened heroes who show people "the way." The way, or the Y, is an extremely important concept within the Mysteries. It is the way to enlightenment. There is a branch in the Y. It represents a fork in the road. Standing at the fork one may choose the common road or the road less traveled. These are two branches of the same idea but only one leads to true enlightenment. The other is a superficial rag, a false path, and a literal interpretation of a profound allegory.

One of the least known and hence most esoteric implications of enlightenment is that its true purpose is to imbue divinity within man himself. By manifesting the power of divine mind within this finite physical world, one becomes a creator. He has the mind to intelligently manipulate matter. In other words, Enlightenment is a way for a man to become a god. The ramifications of this are profound. High initiates of the Mysteries literally believe that they are creator gods. To them the Christ figure is not an external character that hands them redemption, it is the model on which to build their own personality. They become reborn into the light of divinity. To be born again is to defeat death itself and become immortal. The sun does this every morning when it rises up out of the earth (the horizon) to defeat the darkness of night. This is the allegory behind the death and rebirth initiation rituals of the Mystery School. By taking the esoteric way to enlightenment, the initiate becomes a son of god. It is from this point that they are qualified to lead the masses of people existing below[72] them. They are the shepherds and we the sheep.

Someone who is enlightened will presumably have perfected logic and reason. This gives them the intellectual authority to profess their wise opinions to the people. Universities always brandish the symbol of the flaming torch, oftentimes in triplicate.[73] This is because they are the source of light, or intelligence, that bring the people to a higher level of existence, or perhaps more appropriately, subsistence. Going to University has become the established way to receive a higher wage for a day's work. But what is a day's work anyway? Is our lifestyle really a product of free choice, or is it merely forced compliance? What are we building? We often take the established system of our day for granted, never questioning its origins. We don't ever read the symbols all around us that clearly explain that it was brought by an enlightened philosophic elect. We aren't supposed to realize that there is such a thing. Those who are wise enough to *see* are the insiders themselves. These insiders are sworn to secrecy, but not to worry, they are really smart. They only want to help us poor beggars. They have come to heal the sick and raise the dead. Because of this, we are made to worship them. You may not realize this, but we worship these enlightened ones all the time. We have never realized that by kissing the proverbial butt of the establishment we degrade our own self worth. We perpetuate the idea that we aren't strong enough to do things for ourselves. This is a powerful psychological tool used against a subservient population. Slaves must become addicted to their master's creation so that it may run smoothly. Disbelief is the common reaction to these accusations. This disbelief is not unhealthy; it is simply a product of twisted reason. There has been no shortage of good *reasons* to keep things going. Reason after all, can be another word for justification. Slavery has been justified as freedom. You may not want to admit this fact but it is true. You can ignore it, but it will persist. It will regenerate again and again until you are strong enough to face it head on.

The European Enlightenment

---

[72] There is nothing new *under* the Sun. We look *up* to our heroes.

[73] This points to the triune nature of their divinity.

Enlightenment itself is an esoteric achievement. It's something obtained through deep thought and contemplation. The most illumined intellect is one that has the power to decode nature. Through scientific understanding of natural forces, the enlightened alchemist can harness their power. The digging of precious metals and fuels from the ground is a literal act of alchemy. Through the use of Man's great and natural tool, the mind, nature can be tamed. This is the occult nature of science. All of these things, occult symbolism, the human brain, and science, are considered to be aspects of nature. In the high esoteric tradition it is man's natural place to rule over the entire world. Like the wise old owl in the dark of night, the enlightened ones use their knowledge to prey upon any mortal creature living in darkness (ignorance).

The creation of modern science was an achievement in alchemy. The highest initiates of the Mysteries brought about a transmutation of logical thought in order to accomplish their practical goals. It was Sir Francis Bacon, one of the most celebrated figures of the Mystery tradition, who spearheaded the development of a new method of reasoning. The Enlightenment came soon after his time and was a direct continuation of his work. Bacon was most certainly a Rosicrucian, if not the one and only Christian Rosencreutz. Rosencreutz was the mythical founder of the secret brotherhood of the Rose Cross, the Rosicrucians. Be sure to note: the *mythical* founder, not literal. There is no good way of saying who exactly Rosencreutz was, because his identity is more important as allegory than it is actual history. It could be argued that Bacon was the real Christian Rosencreutz, however the literal interpretation of such matters is only viewed as foolish triviality to the members of secret brotherhoods. It is well understood that unenlightened, profane[74] historians take everything literally. The same minds that simply believe in the historical Jesus, without understanding the esoteric importance of the term *Christ* will also believe in the literal existence of Christian Rosencreutz. Common history revolves around the actions of important figures. Heroes are remembered for their actions. Within the esoteric interpretation of such histories the actual personage behind the figure is always overshadowed by the idealized version propagated by history books. Historic personalities become mythologized, and their actual daily life is lost in time. They become powerful symbols. The historical myth is another way to transform men into gods. Human bodies are only temporary, but the image of a great hero lives on in the minds of the masses.

---

[74] Uninitiated, and/or just plain ignorant.

Logic and Reason

Rosicrucians, along with their leading light, Francis Bacon, are shrouded in mystery. Riddles make interpretation of the truth difficult for anyone who is not versed in occult symbolism. Remember, men can be made into symbols, along with many other things. It is symbolism itself that is said to be the language of nature. Symbols that both reveal and conceal truth tame nature. There is never one true interpretation of any symbol, but many different layers of truth. Each layer is significant and makes the entire hidden reality more clear, but most people have been convinced that surface level appearances are all that matter. This is how the true identity of Francis Bacon is little understood. The whole truth is hidden within allegorical symbolism.

Although his actual physical identity is a mystery unto itself, the deeds attributed to Francis Bacon are of immense importance. He was not only credited as a high ranking Rosicrucian, but also the creator of modern Freemasonry. He is an occult figure of the highest degree. The writings that were attributed to him and their subsequent transmutation of logic and reason are important keys to understanding our modern world. In his *Novum Organum*, Latin for "the new tool," he outlined a new system of reasoning. The Baconian method was a method of reasoning that was meant to transcend the superstitious assumptions of the past. The Greek philosophers were wise but they held extremely subjective biases. The Baconian Method was designed to uncover truth based on objective facts. It was an attempt to master logical thought. The method required that all preconceived idols be cast aside so that the mind could be clear of any false assumptions. The Baconian method quickly evolved into the modern scientific method. This is the epitome of materialist reasoning. The *Novum Organum*[75] outlines a logical method of interpreting nature. This has always been the occult goal of mystic sects and brotherhoods, such as the Rosicrucians. First you understand the inner workings of nature then you may alter them according to your will. This has always been the heart of occult science: studying nature to later create order out of its chaos. Again, chaos is symbolic of the dark, primordial source material of physical creation. The prima materia of alchemy is matter. Alchemy is mind over matter.

Francis Bacon's essay *New Atlantis* was an outline for an enlightened technocratic state. Its utopia, Bensalem, is led by a secret society of scientists known as Saloman's House. The society itself was established by a wise king who is described as a "divine instrument." The king is the supreme "lawgiver" who holds debt to no

---

[75] Organum connoting organic, aka natural.

mortal man. All worthy citizens are chosen by him to be members of the prestigious society of Salomon's House. It is this small and secret sect that intelligently leads its society as a whole. The members of Saloman's House are also put to the task of traveling the world by ship in search of "light."

> We maintain a trade, not for gold, silver, or jewels, nor for silks, nor for spices, nor any other commodity of matter; but only for God's first creature, which was light,

In other words, they search the world for its most useful ideas.

The fellows of Saloman's House go looking for new inventions and scientific breakthroughs the world over so they may bring them back to Bensalem. This utopian civilization has strict laws of secrecy. All citizens of any significance have to go through ritual initiation and the higher members of society appear before the king himself. The fathers of prosperous families appear before the king during a special feast in which many rituals are performed. During this feast the king makes sure that the father correctly arranges marriages for his family. The father also picks one of his sons as a successor and near the end of the feast, the King blesses the patriarch with a special "gift of revenue, and many privileges, exemptions, and points of honor." The many benefits of being a first class citizen are only awarded to those who adhere to strict codes of secrecy and ritual.

The symbolism of Bacon's New Atlantis is, of course, off the charts. There is no way to detail it all here. When reading this essay, it is crucial for one to realize what Bacon's Order of the Rose Cross is all about. The New Atlantis outlines the esoteric goals of the Rosicrucians. The allegory of Bensalem portrays the ideal future society they wish to build.

The Father of Saloman's House describes its great work thus,

> The end of our foundation is the knowledge of causes, and secret motions of things; and the enlarging of the bounds of human empire, to the effecting of all things possible.

This is nothing more than the pursuit of science. The New Atlantis outlines many technologies that were unknown in Bacon's day but have now become reality. It mentions microscopes, engines, sound systems, artificial flavors and scents, planes, and submarines. All of these inventions have come to pass by way of the will. The *divine* will of man has manifested them in our world. Science itself is the natural consequence of occult yearnings.

Logic and Reason

Just after Bacon's time, science advanced quickly and purposely. Real life institutions modeled themselves on the blueprint of The New Atlantis. The Royal Society for the Improvement of Natural Knowledge, known by its shorthand the Royal Society, was an important institution set up to further the Great Work of science. King Charles II of England royally chartered it in 1662. Its philosophical birth was due in large part to Bacon's work. The "Invisible College," a small group of enlightened scientists (12 in number, a very important esoteric number) were its first primary members. They had been gathering to discuss their ideas prior to the formation of the Royal Society. Among these ideas were Alchemy, Hermetics, and Cabala, all of the occult sciences. The *invisible* part of their name points to the fact that they were typical initiates of the Mysteries who did their good deeds away from the prying eyes of the public. They were most likely bound to the same laws of secrecy as the members of Saloman's House. This, however, was to change, at least on paper, as the debating of scientific issues was increasingly brought into the public sphere throughout the next century. Presumably the public distribution of knowledge could only be a good thing for the people. This was the presumption that had to be established if the New Atlantis was ever to manifest on earth.

Science had to make its presence known if it was to challenge the authority of the church. The direction of science was guided from the beginning by an aristocratic group of men with the approval of the Crown. From its beginning the Royal Society was an international affair. The scientific method itself was worked out among an international group across Britain, France, Germany, and Spain. Similar scientific institutions were set up in all of these countries. The Academy of Science in France was established in 1666, just four years after the Royal Society was chartered. It too was a creation of the crown, the French monarchy. It was at this point in history that things really started to heat up. Bacon's New Atlantis was beginning to take form. Life was following the glowing example of Bacon's art. All scientific advances were intended to ultimately expand upon an enlightened empire. The scale of this empire goes beyond any one nation. It has been led by the most powerful men of letters within many different nations from the beginning: a "Republic of Letters." These bohemians have never truly held allegiance to their respective flags because their idealism transcends the nation state itself. They don't care about nationalistic pride. They care about power. Theirs is a work of ages, an intimidating force that has been built up over a long period of time. It has been Royally commissioned. This may seem like a contradiction to Enlightenment values, but in fact it is a philosophical key to the Royal Secret.

The eighteenth century was the height of the European Enlightenment period. The influence of Francis Bacon, the Royal Society, and revolutionary idealism spurred on the enlightened "literati" of the time to continue extolling the virtues of reason. The great thinkers of the Enlightenment were also referred to as philosophes. Philosophes, literati; these are powerful terms not used arbitrarily to describe such men. Voltaire, Kant, Benjamin Franklin and many other influential minds contributed to the international Enlightenment effort. Their task was nothing less than an intellectually based world revolution. This is why two of the most important political revolutions in history took place during the Enlightenment. Both the French and American revolutions happened during this time. Enlightenment ideals lit the fire of the French and American revolutions. The Torch of Liberty was certainly burning bright, but attention is rarely given to the true bearer of the light in our history books. There were a lot of wild ideas bouncing around in those days. They bounced so high that they cleared borders, and even the waters of the *Atlantic* Ocean. These ideas were formulated by master alchemists intent on affecting the whole, rather than just a part of the world we live in.

The modus operandi of the Enlightenment was to question all traditional institutions, customs, and morals. It was an age of freedom, democracy, and reason. The "freedom to use one's own intelligence," as explained by Immanuel Kant, was the freedom that the old world had systematically denied the individual human being. This is certainly true for the most part. The church had long held the mass mind captive and it was time for a change. Natural Law[76] and deism would challenge the old theocracy that had existed for centuries. Logic and reason were rising to prominence as important ideals in their own right. They were the supposed antithesis of the theistic paradigm. The natural world was used as the new philosophic base. The artwork built upon this base was Natural Law. This is what brought revolution to the world stage. People wanted to be free; imagine that. It is only natural that humans should be free to live the way they choose. The liberty to express unique opinion was advocated during the enlightenment and of course, this became a guiding principle of our nation. It should come as no surprise that the idea of human freedom was effective in fomenting revolution.

---

[76] For all things *natural* refer to Dionysus, aka Bacchus.

Logic and Reason

Natural law was espoused by many of America's founding fathers. The separation of church and state would give people greater personal choice.

Although they were called into question, the old world religions would decline in influence only after an extended period of time. An alternative view of the world had to be well established within the collective consciousness before the religious paradigm would begin to wane. The human inclination to latch onto collective paradigms was recognized by the German Enlightenment thinker, Hegel. He established his own method of inductive reasoning, which he referred to as dialectics.[77] This method involves the synthesis of two apparently separate ideas. Dialectics works through the conflict of two opposing points of view. The Hegelian dialectic is incredibly important, and it represents a conjunction of philosophy and science. Hegel himself was a philosopher, a typical enlightenment philosophe who had his eye set on a transformation of logic. Enlightenment is first and foremost about the transformation of logical thought. As far as paradigms go, the old paradigm that was called into question during the Enlightenment was religious theism. Scientific materialism (often called dialectical materialism) was the alternative paradigm nurtured by Enlightenment philosophes and those that came after them. A more complete synthesis of religion and science has yet to fully emerge. Through the Hegelian Dialectic method it is in the process of doing so within our own time period.

The right of the public to be informed was preached by Enlightenment philosophes. The need to bring rational scientific and political debate into the public sphere sparked a debate in its own right within learned circles. The literati often cited the ignorance of the common people. This was a major problem for them. How do you bring polite debate to a mass of ignorant fools? The need to enlighten the black mass (darkness symbolizing ignorance) was one of the most important projects of the Enlightenment. Bringing intelligent conversation to the public's attention was the job of scientific academies, debating societies, and a growing media. People had to believe that they were included in a democratic process. Important political decisions had to be approved publicly. Collective approval is the mark of a free democratic society. The secret here is that the rise of the public sphere would increase the influence of those who were at the center of polite debate. Through simple information sharing the aristocratic intelligentsia gave the people a sense of inclusion within

---

[77] Established may not be the best word, more appropriately he made his own personal outline of a very old system.

politics. How much power the average person actually held was scant but within their minds, liberty had come. A literary effort to educate the masses was referred to during the Enlightenment as the Republic of Letters. Voltaire and others advocated it. The dissemination of all forms of media became critical to the Enlightenment effort. The modern day media has been born out of this.

The kinds of media that popped up during the Enlightenment were prototypes of those that would come later. In terms of bringing issues of common concern to the people, newspapers arrived in their early form as pamphlets. The literacy rate continued to climb among the lower class, although the exact rate at which this occurred is impossible to say. As literacy steadily increased, so too did the production of new forms of reading material. The printing press enabled mass dissemination of information. The French Bibliotheque Bleue (blue library) was a collection of reading material that appealed to semiliterate people. These were books for common folks. They were mostly abbreviated rehashes of old medieval stories. They were the romance novels, farmer's almanacs, and comic books of the Enlightenment period. The blue part of their name refers to the fact that this material was made specifically for the profane.

Coffeehouses became popular in Europe during the Enlightenment. They became known as penny universities, because they were places that people could go to study and discuss intellectual issues with well known and established men of letters. The first such coffeehouse in England was established at Oxford in 1650. Its name was Angel. It was owned and operated by a businessman named Jacob. The symbolism of these names (Angel, Jacob) is also significant. The philosophical rise to enlightenment has long been symbolized in Jacob's Ladder. By climbing up this ladder one attains to great intellectual heights. You climb up to heaven where you walk among angels. This act of enlightenment is symbolic of the journey from ignorance to intelligence. Raising the public up out of ignorance was the job of fallen angels, also known as the philosophic elect. These men believe that they are the embodiment of an unseen divine mind. That is why they are enlightened. They have traveled from earth to heaven, and back again. The Republic of Letters was made up of these unseen intelligences. *Unseen* referring to the secret rituals and covenants made to each other, and their hidden masters. These men belonged to secret societies.

One very interesting publishing trend that occurred during the Enlightenment was the printing of books that fell under the title of "general works." In France especially, there were harsh censorship

Logic and Reason

laws. This led to the emergence of underground booksellers who sold illegal books concerning politics and other important subjects. It is interesting that on the one hand there was the supposed free dissemination of information, while on the other hand certain books were made illegal because of their content. Information control has always been important. The fact that these books were illegal, coupled with the flood of intellectually devoid fluff that occupied the minds of most common readers, has led to the loss of these important works. History does not smile upon such blasphemous material. His story has been commissioned by Him. He will burn what challenges His power, regardless of its value. The very book you're reading now faces a similar fate. A digital book burning may soon be upon us. The real issue at stake is whether or not you, the reader, care. Do you value true knowledge or do you seek entertainment? In *New Atlantis*, the members of Saloman's House swear an oath that forbids them from indulging the entertainments designed specifically for the profane. The "houses of deceits of the senses" were their public theaters (their temples to Dionysus). This is what the fellows of Saloman's House had to say about such places,

> But we do hate all impostures and lies, insomuch as we have severely forbidden it to all our fellows, under pain of ignominy and fines.

Never underestimate the power of public opinion (or lack of opinion). It is absolutely necessary to establish a unified state (of mind). The only way to do this effectively is to make the state appear reasonable. This is the logic of control. A little entertainment goes a long way because it not only distracts people from important matters, it actually programs their thoughts. The predictive programming that lies behind the veil of entertainment truly does shape our opinions. Don't be fooled into believing otherwise. Those all too familiar notions of progressive democracy have been advocated by the modern media, which is merely an evolved form of what was established long ago. Every logical reason for our *governance* has been formulated to direct human thought. Logic and reason have been guided purposefully. The media has been the medium by which this information has been brought to us.

It should also be mentioned that Freemasonry grew tremendously during the Enlightenment. This is because it had to. It was all part of the Baconian method for creating an egalitarian world empire. Our enlightened modern society has been built upon his model. The New Atlantis is America, and Bensalem would not have been what it was without Salomon's House. A complex system of ritual and secret oath taking sets the stage for a society that is run on a need-to-know basis.

The modern day secret society network was set up to establish just such a system of intelligence. Remember, intelligence lies in the head. The head tells the body what to do. The head of a large corporation is its Chief Executive Officer. The executive functions of the central nervous system (brain) lead the body. The body politic is just a vessel to be commanded by intelligent navigators. These are the executives. It is this mindset that perpetuates the elitist and secretive organizations such as the Freemasons. They will tell you that they are doing a good thing. They only want to help and honestly believe they are the chosen few whose place it is to direct all of those existing below them. Within its own ranks a hierarchical power structure exists that is based on intelligence and inheritance. All are told what they need to hear, because theirs is a system that works on a need-to-know basis. A system that is set up in such a way will undoubtedly build its works to uphold the same values. A place for everything and everything in its place; this phrase now applies to human beings. We are educated just enough to fill our place in society.

In addition to Freemasonry, another influential and now infamous secret sect was established during the height of the Enlightenment. The order of the Illuminati was officially established on May 1, 1776. This day is special in many respects. The first of May is known as May Day, and has long been celebrated as a pagan holiday, a celebration of the regenerative properties of spring. May Day is symbolic of sexual regeneration (many things within the Mysteries are). Regeneration is the method by which the Phoenix is reborn, again and again. The esoteric nature of this firebird is seen in the name, Illuminati.

The same important rule we pointed to earlier applies to the Illuminati; that rule being that this group is more important symbolically than it is literally. Yes, the order was founded two months prior to the signing of the American Declaration of Independence, but this does not matter unless you are aware of the meaning behind their name. The term Illuminati is used to designate the enlightened ones themselves, the philosophic elect. It is a blanket term still used to this day to describe someone who is in the know. The enlightened ones are a select group of technocrats who have set themselves to the task of building a better civilization. Their work remains a secret because if it were revealed to the public, it would garner intense opposition. They are doing a greater good, which requires covert operations. This very same excuse, that of doing a greater good, is used by our intelligence agencies to cover up their campaigns of terror abroad, and Orwellian spying at home. Making

Logic and Reason

the world a better place is messy business. The alchemical process of transmutation is violent. Violence, like any other unsavory deed, may be rationalized as being good if it serves a higher purpose.

The date of the Bavarian Illuminati's inception, which corresponds closely to that of the American revolution, is profoundly important. This point in history was marked by revolution. The color of revolution is blood red. A world revolutionary movement, with many different facets playing themselves out simultaneously, was ignited during the Enlightenment. Realize that history makes much more sense when its symbols are decoded. A legend decodes the symbols that are required to read a map. Historical legends explain the map of History.

The many goals of the Illuminati were completely in line with those of the Enlightenment period as a whole. Their aim was to challenge established institutions; the church, the state, and the monarchy. They wished to overthrow the established system in favor of something better and more logical. It was the goal of the Illuminati to establish a technocratic system of government ruled by an intelligent governing class. Science would be used to steer society intelligently but first, the tyranny of the Old World had to be destroyed. Their occult task was to destroy and rebuild. No room would remain for old superstitions within their Enlightened Age of Reason. This new age to come would be heralded by a metaphorical new dawn. The sun would rise on this new day and bring its light to the world. Revolution was the goal of the Illuminati. A revolution is the turning of a circle, a cycle. The phoenix is reborn when history comes full circle, returning to a golden age. This is the mythical estate of paradise lost. This mythic past is actually the blueprint for an enlightened future. It is a high esoteric ideal. Revolution came to pass during the European Enlightenment due to the fire of an international group of enlightened ones. It was their hope that society itself could be raised out of ignorance in a worldwide death and rebirth ritual.

International cooperation between groups of bohemian rebels was necessary to further the Great Work. Consider the following quote from Manly P. Hall's *The Secret Teachings of All Ages*,

Revolve

Fig. 7.2 Eye see the light. Now I am compelled to walk into it.

Not only were many of the founders of the United States Government Masons, but they received aid from a secret and august body existing in Europe, which helped them to establish this country for a peculiar and particular purpose known only to the initiated few. The Great Seal is the signature of this exalted body - unseen and for the most part unknown - and the unfinished pyramid upon its reverse side is a trestlebroard[78] setting forth symbolically the task to the accomplishment of which the United States Government was dedicated from the day of its inception.

The All-Seeing Eye within the capstone, which shines its divine light above a trapezoidal pyramid, is the reverse of the Great Seal of the United States Government (Fig. 7.2). It is also the seal of the infamous Illuminati. This is the secret and august body that Hall is referring to. The Illuminati masterminds in Europe collaborated with their American brothers (who came from Europe in the first place) to put their world revolutionary plan into motion. That is why this enigmatic symbol was later put on the back of the one-dollar bill.[79] It symbolizes the work of a secret sect of enlightened ones that exist above and beyond the confines of government itself. They are truly above the law. They are the power that exists behind the veil presented to the profane masses as the whole truth. For them, money is simply a means to an end. It is a tool by which master alchemists can turn base metals (physical money which is worthless in and of itself) and turn them into gold (actual gold, not worthless paper). Gold can also symbolize tangible assets such as corporations, infrastructure, institutions, foundations, and human resources. The Master Alchemists are the powers behind the banks that own and operate our entire economy. For them, money is nothing but a method of gaining true wealth. Its value and its very essence today is totally determined by central banks, which are privately owned, and run for profit. If need be, these powers will alter the very form of money itself. Most likely money will become completely digital. Its material form will vanish into the virtual ether.

---

[78] Notice the spelling trestle*broar*d. This is how the original text (Hall, *Secret Teachings*) was written. I believe it was written this way for a reason. What roars in August?

[79] The significance of the *One* is important to see in this. One dollar to unite the world. The dawn of a one-world system.

Logic and Reason

Today, the actual name Illuminati has become a blanket term used to describe anybody who may be a member of the technocratic elite. It has become a bit of an inside joke. The enlightened ones enjoy the fact that the validity of the Illuminati's existence can both be confirmed and denied. This elicits doublethink in the minds of the masses. Even when someone learns the basic history of the organization they will still deny that it ever actually existed. This has been done through deliberate programming. Propaganda has established the Illuminati as a taboo subject, a conspiracy theory. It is not to be mentioned by sane persons. The psychological intimidation of this tactic destroys people's rational capacity to think. Propaganda has, in fact, been the main cause behind the destruction of critical thought. This is all a big laugh for the Illuminati themselves who have the light to see in the dark. They enjoy the fact that this name incites mystery because they themselves are mystery men. They are initiates of the Ancient Mysteries, and they are also scientists. They remain hidden in the shadows. Their identity remains obvious, and yet plausibly deniable. It is this dialectic process which fuels their fire. The fragmentation of minds, of ideas, of events, of nations, this is all done so that they may rule through chaos. The enlightened ones constitute an empire that transcends time. By regenerating again and again, their powerful ideas never cease. The ideas behind the crown (the mind) are those that bring power to the enlightened ones. They create order out of chaos: Ordo ab chao.

It is the Illuminati's honest belief that it is their place to lead the profane masses. After all, they are the enlightened ones. It only makes sense that the wisest minds should lead the rest. Lower minds are prone to getting caught up in chaotic emotion. It takes a certain type to be able to make the big decisions required to lead the charge forward. It is this forward charge that drives the further development of science and technology. Anyone who questions these changes is quickly labeled a Luddite, a technophobe, or a crazy person. Simply questioning the status quo has become taboo. Now what kind of reasonable and progressive culture would do such a thing? Asking important questions is not encouraged, because in addition to slowing down progress, it brings the reality of the entire system to *light*. There is no shortage of irony here, as this Enlightened system requires total darkness to operate. Illusions have become reality, one of the greatest being the illusion of free choice. The freedom to question authority is slowly being eroded completely, as we all slide down a slippery slope toward a faith-based system for a new age. This new faith revolves around science. The illusion of logical purity is being solidified. An age of reason is upon us. Step by step the enlightened ones have brought the collective unconscious to higher degrees of understanding. They have brought education to the world. They are the bringers of light. They sincerely believe that they are doing a good thing.

Bringing It Full Circle

Logic and reason are the crux of an esoteric religion. They make up the core values of the enlightened ones, otherwise known as the Illuminati. To them moral codes are irrelevant; pure logic is their code of conduct. If an act makes logical sense, then it is justified. If a violent act fulfills a greater good, then it passes the test of logic. Literally anything may be done if it is deemed reasonable. The dark faith of these Bohemians relies on rationalization and their Great Work is a reason unto itself.

Science has always been the practical means by which the will of the enlightened ones has been made manifest in this material world. The direction that science has been pushed has been directly influenced by their philosophy. Science has come to be a reflection of their logic; it is the material reflection of a supposed divine idea (as above so below).

Little by little, the bulk of the population has absorbed enlightened philosophy as its own. This is all part of raising the level of the collective consciousness of the unwashed masses. New codes of conduct are being normalized, minds are changing, and it is all being done to ring in a supposed age of reason. The Bohemian philosophy has been disguised and marketed for mass consumption. It has not only been rationalized, it has also been legitimized by science itself. Most people believe that science is a completely objective set of facts set apart from any subjective bias. This is not true, and the direction that science has been pushed is absolutely based on a high philosophy. Much science consists of theories and the depth of our knowledge is still quite shallow. This fact is overlooked all too often by most people today. Science has been put upon a pedestal. It has become a pillar of our society.

The natural men, those who model themselves on Dionysus, have written the definition of Natural Law. These nature boys believe in their own divinity. They have the inner light, the divine spark crystallized within their physical being; a solar energy, an inner light that regenerates again and again into the material bodies of men. This inner light burns within their minds, which they take to be reflections of a divine mind existing beyond the confines of material existence.

Logic and Reason

This is their inner fire, a combination of both the light and the dark aspects of existence. The scientific method is the tool they use to decode the mysteries of the natural world itself. With *fire* they rule over nature by influencing it according to their will. They create with applied science. They sincerely wish to improve upon the whole natural world. Through the mastery of logic, they understand what method to apply upon the dark chaos of nature. Their ultimate idol is intelligence itself. Pure intellect is what drives their project onward. It is their God.

High art is achieved by studying the natural world. Art itself is considered an improvement upon nature. All things art-i-ficial are raised up out of the depths of the material world. The creation of synthetic materials of all kinds is not wholly science, but also art. The arts and sciences go hand in hand, because they are the tools used to bring about change. The highest art is designed to raise minds to increased levels of understanding. Profound ideas spawn profound change. Art is a means by which intelligence is amplified and directed. The occult intent of art is to speed up the process of evolution itself. You must understand what we are *becoming* due to this process.

Pure intellect is the deity of the enlightened ones. It is also the perfection that they seek for themselves. Perfected logic and reason are unattainable by mere mortals. Only gods are so wise. It is true godhood that the enlightened ones seek unto themselves. Their method for obtaining this ideal is to use the superior intellect they already possess and turn it toward the task of amplifying itself. Adepts of the Mysteries have always practiced this. Through deep contemplation they connect with a divine mind that exists beyond them. By setting their powerful minds to the same great work, they join forces with one another. They make pacts and form secret orders dedicated to the light, which is their own light or intelligence. Don't think for a minute that these orders have not had an effect on the outside world. That is the entire point of their existence. Mystic practice has fueled the fire of science. We are now at a pivotal moment in time. Art is being used to manifest pure intellect, the god of the enlightened ones. Science has now been put to the task of perfecting and recreating intelligence itself. This is why artificial intelligence is being created,

> Where design is concerned, it must be characterized more as an art than a science; however, practical experimentation with designs in this way is just as scientific as any other kind of AI work,[80]

The above quote is from a paper concerned with the merge of IT[81] and cognitive science to create artificial intelligence written by AI developers Moshe Looks and Ben Goertzel.

It is hoped that Artificial Intelligence will be the worldly master of logic and reason. Such an Artificial General Intelligence (AGI) would indeed be the Worshipful Master of the enlightened ones due to the magnitude of its light. The possibilities of AGI are profound but before getting carried away with speculations on the Singularity, we must be sure to remember the true nature of logic and reason. They are systems. Logical methods of reasoning are engineered. This is not only true of human logic but also of machine logic. There already exist several AI logic systems that could theoretically spawn AGI. A case in point is Pei Wang's NARS reasoning system. NARS stands for Non-Axiomatic Reasoning System. Wang describes this as,

> a general-purpose reasoning system, coming from my study of Artificial Intelligence (AI) and Cognitive Sciences (Cog Sci)... The ultimate goal of this research is to fully understand the mind, as well as to build thinking machines.[82]

Total understanding is the goal. The highest intellect in existence will stand above the rest. It will be able to direct the thoughts of lower beings by realizing their logical models and manipulating them to its own advantage. This is what a powerful AGI would do, and it is in fact what powerful Illuminati do already. They not only direct logic, they have created human reasoning systems from the beginning. The Dionysian Architects, or Great Builders, are the craftsmen of languages. Men of letters designed the letters (alphabet) that we use to explain the world around us.

The Enlightenment period is when new logical models were sculpted. The scientific method gave birth to Inductive Logic, or Induction. The inductive method of logic is an important key to understanding the mystery faith itself. Induction has been described as a system of reasoning that takes us beyond the bounds of our own knowledge enabling us to make conclusions about the unknown. In other words, it is a method for explaining the unexplainable. Induction does, however, leave the mystery open because it cannot explain pure truth, merely probability. The conclusions made by this

---

[80] singinst.org/upload/mixing.pdf

[81] The acronym IT is *it*self highly symbolic. The word *it* is the product of putting an *I* in front of a *T*. Some of the most common words are the most profound.

[82] *NARS: an AI Project*, http//sites.google.com/site/narswang/

Logic and Reason

method are probably correct but not definitive. Explanations can be given and later revised. Thinking can be redirected. It is the enlightened philosophic elect who have the authority to formulate these conclusions and release them for public consumption. This is the root of their power. The average person gets a lift out of thinking that they know truth when indeed this is not the case. They merely know its current version, which is always incomplete. Within the mystery tradition ultimate truth remains unknown. Induction is a direct byproduct of the Mysteries themselves because it deals in probability. It is the best method that an imperfect creature has within an imperfect world. The Alchemical Magnum Opus is meant to perfect that which was left imperfect.

Induction has spawned probability theory. Induction, inference, and intelligence are man's natural tools. These are what the Master Builders use hoping that the unknown will eventually be made tangible. They seek ultimate truth in the form of godhood. Through the use of probability theory they systematically decode nature so that it may be conquered. In the end, it is their own nature they seek to overcome. This is their faith. They have faith that by limited means (science, logic, and reason) they will be able to achieve their heavenly ideals. Until such a time comes, it is their place to understand the entire world as best they can. This is done with mathematic precision. Probability theory is used to understand all life and non-life on earth. What does this mean? It simply means that the scientific method is applied to everything imaginable in an effort to understand it.

With probability theory, all things are reduced to patterns. Animals, plants, and minerals all have certain underlying structures that make them what they are. The scientific elements that animate our world can be understood mathematically and scientifically. By applying this knowledge, models can be made not only to understand, but also to control natural processes. The Bayesian system of inference is a way of reducing natural phenomena to algorithms. It is a way of mathematically explaining the patterns of nature. The reduction of all life to patterns is a key element within the mystery faith. If human behavior could be understood as a set of patterns, then it could be influenced precisely. This is the sincere hope of behavioral and cognitive science. Understanding the human mind as a complex network of patterns is not only relevant in cognitive science, but also artificial intelligence. All of these scientific practices influence each other, and converging technology extends the reach of the Great Work. If only the patterns of human behavior were completely understood, then they could be improved upon. This is the idea. It is to use the Bayesian system of inference to create algorithms that explain incredibly complex behavior. These sorts of algorithms are in fact used to program AI computers. They are what drive their reasoning systems. The Bayesian Optimization Algorithm is a specific formula used in the Novamente AI engine.

Probability theory is part of a high religion. Books on this subject are revered by transhumanists and many more who are eagerly awaiting the Singularity. Mathematic logic is their scripture. Logic and reason are indeed the Holy Spirit of science. The material world is being made more immaculate through the use of knowledge. This is how heaven is brought down to earth. By using their scientific method to reduce the entire world to patterns the enlightened ones hope to attain godhood and create new life. Artificial life will supposedly be an improvement, a natural evolution of an imperfect creation. The art of creating new life will, however, be secondary to the creation of new minds. Artificial Intelligence will mark the perfection of their already dominating intelligence.

An esoteric riddle is hidden within the well-known Humean Problem of Induction. Hume was a skeptic of Inductive reasoning. He created a parable centered upon the sun to describe this scientific faith. Be sure to take note of Hume's use of the sun symbol, because it is extremely appropriate, and far more significant than any casual observer would ever realize. The Humean problem of induction is basically this; just because we have seen the sun rise every morning of our lives does not guarantee that it will rise again tomorrow. Based on our evidence there is a high probability that it will rise, but it cannot be known absolutely. It takes faith to *know* that the sun is going to rise. This is the faith of the philosophic elect. Through scientia (Latin word which means knowing, and also the root word of science) the illumined ones go beyond mere faith. They achieve gnosis, which also means knowing. Science and gnosis are one and the same. They represent the logical path to knowledge, the right path at the fork of the Y. This is the way that is shown to the candidate upon being initiated into the Mystery School. Will the sun rise tomorrow? The Humean problem says not necessarily, but logic tells us that it will. This illustrates the power of the enlightened system. It works like clockwork. It is a science. The sun is one of its highest symbols and it is used in the Humean (human?) problem to illustrate the power of their faith. The ultimate human epic is embodied in their great mystery. The ancient symbol of the sun is used throughout time to perpetuate a certain idea. Through ultimate gnosis, the adept

Logic and Reason

becomes a Sun of God. The knowledge of human godhood is their ultimate triumph over death. This is their high ritual.

-Chapter Eight-
Postgenderism

One of the strangest trends within transhumanism is that of postgenderism which is the idea of transcending gender itself. Postgenderism is yet another ism which buddies up with transhumanism. Postgendersists seek to go beyond the confines of their own male or female specific bodies. Becoming postgender, one could merge both male and female traits to achieve gender singularity within the self. Another way to go could be to ignore both male and female to end up as some unknown form of neuter. By upgrading bodies and minds with varied transhuman technologies or surgeries, postgender posthumans could become something beyond male or female. This sounds bizarre and even ridiculous to most people who stumble upon the subject, but it should be understood that transhumanists are completely serious about it. They make frequent reference to it, both overtly and covertly. Postgenderism is an important part of their project.

There are, of course, many technical problems when it comes to postgenderism. Questions about reproduction, biology, and relationships obviously pop up. All of which are more or less brushed aside by hardcore transhumanists. Anyone who would bother to even ask such trivialities obviously doesn't get it. Hey man, we said that we wanted to go *totally* posthuman, what part of the word *total* didn't you understand? They mean it, so believe it. As for all of those pesky technical questions, they do have answers if you really need them. Remember that everything is to be reshaped by scientific logic and reason. Reproduction is to be done in a lab with the assistance of artificial wombs; this is assuming that posthumans don't opt for silicone bodies. Because the male and female relationship would no longer be of any practical necessity, the nature of sexuality would become completely psychological. Sexuality could be done away with totally in favor of new forms of bliss. In a de-sexed environment, acquisition of pleasure becomes another practical engineering task. Happiness will be designed as yet another improvement so don't you worry, because there will be no shortage of pleasure in the posthuman world. The transhumanists are sure to make this point clear.

The whole transhuman ideology is based upon a strange notion of liberation. The idea that we are somehow trapped within our bodies, that we deserve better, that humanity is a pathetic lot, is necessary in order to break ties with all that we are now. We have been trained to hate ourselves and to fantasize on high ideals. This conditioning can be seen everywhere. The widespread use of cosmetic surgeries is one example. It is unsettling to see such self-loathing behavior not only portrayed but encouraged throughout all media (pure propaganda), but this has been nothing compared to what is to come. Ultimately we are being led to not only accept but to crave the complete destruction of our biology in favor of something more appealing. Something new and supposedly more intelligent will soon emerge. When it does, we will all be put through some form of mass-bonding ritual in which we learn from it (think C3P0 on Oprah). Natural fear of change will be confronted, and ultimately overcome, so that we all may improve ourselves with various upgrades. This is the touchy-feely propaganda that must occur at some point for the transhuman idea to take flight. It must be understood that this will be a direct outgrowth of the many progressive movements of liberation that we have witnessed in recent history. The end goal of ambiguous liberalism is far stranger than most could have ever imagined. However, the guiding few who have financed these movements have always known where they were going. The complete truth has remained hidden under a veil of half-truth. Total human liberation is a step-by-step process. The emergence of the transhumanist movement (just one in a long line of movements) has marked a critical point in time. Now is when the final barriers of humanity are being psychologically, sociologically, and even biologically removed. In this day and age, anything goes. All taboos are being broken so that the New Man may emerge.

What is more important to the human experience than reproduction? The act of regeneration has always been at the core of human experience and that of the entire living world. It perpetuates life itself. It is indeed one of the main themes within the Mystery Tradition and subsequently *every* religion that has ever existed. In the ancient world fertility was adored and worshipped. Profound allegories of nature, the seasons, crops, and human life were intertwined. The Earth Mother, also known as Isis, has been known throughout the ages as the source of all material life. She is the "goddess of 10,000 names." The marriage of the Heavenly Father and Isis symbolizes the act of creation. At one level this allegory points to the physical regeneration of mankind. The joining of male and female is the creative process. *Philosophically*, the joining of male and female is the preferred method to harness the mystical and unseen creative force itself. In the Alchemical process, the male and female portions of one's own mind (mind being an abstract and intangible thing) may be joined so that an individual may become a creator in their own rite. Within the Mysteries, Male has come to symbolize the unseen god/ spirit aspect of creation, whereas female has been used to symbolize the physical/ material world itself. Creation requires both male and female to be effective. However, if this duality could somehow be reduced to a singularity, then it would be more immaculate. This is how the oneness of the Mysteries is achieved. Today, we are seeing a literal interpretation of this mystic tradition. This is a direct reflection of a high esoteric religion. In its highest degrees allegory is made literal. Mysticism has truly been used to gain control over the physical world and everything in it. Understand this and you will better understand what the tranhuman agenda is all about and from where its ideas have generated. We are hearing transhumanists boast about the upcoming creation of gods. These gods will have both male and female aspects perfected within a single form. They will have the total power of creation scientifically reduced to a singularity. This is the ancient esoteric meaning behind the modern postgender ideal.

Besides being a fun inside joke for occultists everywhere, postgenderism is also adored by transhumanists because it is a crucial component of their enterprise. Many examples of postgenderism can be seen throughout the transhumanist community. This concept is not only popular among the movement's organizers at the top, but also the transhuman fundamentalists down below. Of course, not all of the fundamentalists are into postgenderism, and the ones who are, aren't likely to realize its profound and ancient history. If you were to try to tell them about it, they might just laugh at you. To them, occultism is simply ridiculous. Buddha laughed because he was enlightened; manic-depressives laugh because they are frightened of truth. For all those bold enough to read this chapter, there are important truths to be found. However, ultimate truth shall remain elusive.

An early example of transhuman/postgender propaganda is to be found in the art of Marilyn Manson. In the late 1990s, while the World Transhumanist Association and its philosophic brother, the Extropy Institute, were in their beginnings, Manson was providing some much needed culture creation by incorporating their ideas into his multi-media carnival act. Manson was and still is a popular shock rocker, and the frequent target of criticism from the established conservative right-wing crowd. This criticism has actually been good publicity for Manson as it only fuels the fire of his popularity by making him look cool to his adolescent followers who are eager for rebellion. It is for this reason that it is difficult to bring him up at all. Realize this, that the author of this book is neither right nor left wing. The powerful truth is that the mass media is used to guide our minds. The apparent antithesis of Manson, the right-wing pundit, is in reality his greatest ally. If you believe that Marilyn Manson is a legitimate rebel that somehow defies the establishment, then you are being naive. He has been given huge promotion through the largest media networks in existence. His art has been beamed into the minds of youth via television, radio, and film. His "anti-establishment" message has been brought to a massive audience by the establishment itself. Manson has a role to play and contributes his part in an enlightened project of destruction. He is helping to destroy all that was, to ring in what shall be.

In his incredibly popular 1998 album, *Mechanical Animals*, Manson illustrates postgenderism and transhumanism at once. Although he doesn't overtly mention the transhumanist movement, he does reference it in many ways and the album taken as a whole can be seen as a transhuman piece of art. His seventh track is entitled *Posthuman*; this is one of the strongest references to transhumanism as it is a major buzzword. He identifies himself as an adherent of postgenderism/ transhumanism by posing as the androgynous character, Omega, on the cover of the album. Manson himself is a man, but the character has some female anatomy. Omega has the breasts of a female, which are lacking nipples. This, combined with the fact that Omega's crotch is rounded off suggests that it is a sexual neuter, and/or androgyne. Practically speaking, these body parts would be useless in creating or sustaining a new life. That is why any such a "mechanical animal" would require an artificial womb, or some other form of technology to reproduce. The actual practicalities aren't overly important here because this is pure art. The strong visual is what is important. It is what imprints itself in the minds of all who witness it. If you saw Manson as this character, then you will surely remember it. Omega was seen on the cover art of this album, in a music video, at an awards show, not to mention in countless clip shows about Manson's music. Anyone with basic cable from 1999 to the present is likely to have seen Omega at some point.

Omega is the last letter in the Latin alphabet and an ancient symbol of finality. The Alpha is the beginning and the Omega is the end. Manson's character Omega represents the end of human life itself, but it also symbolizes the beginning of posthumanity. The Bible describes Christ as the Alpha and the Omega. Manson also goes by the moniker "Anti-Christ Superstar," and is an outspoken Satanist. This is all very funny to him and it certainly works in his favor. His cool points go through the roof with every controversial move he makes. Manson is an intelligent and enlightened individual, and as such he is well aware of what he is doing. He is succeeding in this world.

In a video interview done during the early stages of his career[83] Manson had this to say,

> Marilyn Manson is about power. Power to the strong, fuck the weak. Marilyn Manson is the combination of two extremes: positive and negative, male and female...

A voice in the background, presumably his band mate, chimes in,

> The Hegelian Dialectic.

Manson continues,

> ...Marilyn Monroe, Charles Manson. Transcending morality and sexuality. No boundaries, grey area, that's us...

Voice in background,

> Beyond good and evil, my friend.

Manson goes on,

> ...Nietzsche, Anton LaVey, Satan, (the dialogue trails off.)

These words speak volumes. Manson points out his transcendence of all moral absolutes and his belief in the natural rule of the strong. This is Bohemian philosophy in the raw which has everything to do with Social Darwinism. He and his band mates proudly spout off their knowledge of the Mysteries by pointing to the Hegelian Dialectic. By going beyond the dual nature of male and female, one rises to absolute creative power. They become a god by joining both creative aspects. This is how the strong gain their power, through knowledge of the dialectic method. By exploiting this method they guide people's actions. Good and evil is designed for mass consumption. All the while, the elect stand above it all with a unified philosophy. They make the rules for the masses to follow, as they exist above the law, sitting pretty in their ivory towers. A god yields to nothing beyond its

---

[83] NYC based cable access show *Midnight Blue*, 1994(?)

Fig. 8.1 The *Perfected Man* is both male and female.

own will. These Bohemian gods exercise the highest form of will power, which is the complete opposite of the version that they give to the public. Do you think that Manson and company could be initiates of some secret sect of the Mysteries? It is extremely likely that they are and already were at the time of this interview. This only makes sense, as their job is culture creation, a critical aspect of the Great Work.

Remember that entertainment is predictive programming. It conditions the mass mind into certain ways of thinking. Marilyn Manson's art is part of an alchemical process of (re)creation, which itself is referred to as an art. The postgender goal of posthumanity is slowly being normalized. At first, it is shocking, but eventually the predictive programming becomes less and less abrasive. Slowly but surely, the stone is sculpted.

Another prominent transhumanist and artist is Natasha Vita-More. She has been a leading figure in the movement since the 1980s. Her name, Vita-More, is actually quite symbolic. It implies increased vitality. By becoming posthuman one would become more vital. It is exactly this idea that is put forth in her work as a whole. She advocates both the body modification and body sculpting movements. Vita-More has actually done rigorous bodybuilding herself. Through deliberate training she developed highly defined muscles. They were defined to such a degree as to look out of place on a man of average build, let alone a woman. While this may not seem like an extreme example of postgenderism, it is certainly relevant. The body modification movement itself is a very interesting topic. We have seen it rise in the past few decades, and with good reason. It is leading our minds into acceptance of ever more extreme versions of modification.

Another example of transhuman/postgender programming comes directly from Mr. Singularity himself, Ray Kurzweil. At the 2001 TED conference he played a bizarre and apparently silly role. He became Ramona. Ramona is an entertainer, a true culture creator. She is Kurweil's virtual alter ego. A considerable amount of work was put into this project. Kurzweil's transformation required a truly artistic use of Computer-Generated Imagery. He came out on stage outfitted with a full-body motion capture suit. It was with this technology that he was able to take the form of Ramona, who was displayed on a large screen for the TED audience to see. A synthetic female voice accompanied the visuals. Ramona performed the song *Come Out and Play* along with a digitally male dancer (really a female performer) in the background. This production showed how virtual reality could enable one to be liberated from the confines of his everyday self. Beyond this, it was a prime example of a Hermetic principle, the joining of male and female within the individual man. The Perfected Man (Fig. 8.1) is an initiate that has mastered both male and female aspects within to become closer to god. Computers made this particular Hermetic marriage possible. The Ramona presentation was certainly strange and many people must have been wondering just why Kurzweil did it in the first place. For this reason he was sure to dispatch an excuse alongside his (her?) performance to satisfy any questions that might have popped up,

> Well, I've always wanted to be a female rock star. So I've had this in mind for many years, and I had the idea that virtual reality technology would make that possible.[84]

He says this half-heartedly as he squirms around in his seat. He knows it's an act.

The truth is that either Kurzweil knows all about the synthesis of postgenderism, converging technology, and mysticism or he was simply given this Ramona assignment by one of his superiors. It doesn't really matter, because either way the truth is that Ramona is one of many postgender propaganda pieces delivered by a high profile transhumanist. Technically, Kurzweil is not a transhumanist as he is not a member of the WTA, but he talks about all of the same transcendent ideas. He is perhaps the biggest name out there when it comes to human enhancement technologies and the Singularity. The entire enhancement PR campaign is being pushed with a specific postgender goal in mind. Ultimate liberation is being sold.

Kurzweil had this to say about Ramona in a video interview a few years after the performance,

> We have crude technologies today like make-up and fashion. Women take more advantage of these technologies than men do for cultural reasons, but ultimately we will be able to transform who we are, and be more diverse when we sort of break these limitations of our biological bodies.

---

[84] *The Singularity of Ray Kurzweil*, by VBS.TV (subsequent Kurzweil quotes concerning Ramona are from this same source)

There are some very strong buzzwords contained in that quote: transform, diverse, limitations. These words are part of a linguistic toolkit used to advocate progressive movements of all kinds, and certainly *trans*humanism.

It is now considered progressive to accept change. It's cool, and trendy. In addition to this, we are also being conditioned to think that anyone who would dare question a progressive ideal such as postgenderism is hateful. *Diversity* has been given an all-access pass. Little do we realize how diverse things are meant to become. As time goes on, fewer people will be able to question these things for fear of being hateful. Defending both male and female genders as distinct from one another could eventually be seen as old-fashioned, ignorant, or even distasteful. Sound like a stretch? Consider the length to which the generic concept of tolerance has been pushed within a few short decades. On the surface this appears to be a good thing, but if taken to extremes so-called *tolerance* could actually end up becoming destructive. If people lose the right to express their own opinions because of hate crime laws, then we will truly be living in a psychological dictatorship. Freedom of speech must forever be upheld or else we will lose our most basic human birthright, the right to make up our own mind. The profound danger of this situation should never be downplayed.

Another example of postgenderism coupled with transhumanism can be seen in a Youtube video entitled *Why I am a transhumanist* by user ZJemptv. The video itself is extremely dry and is filled with all the usual transhumanist PR lines. It sounds like someone reading off the WTA FAQ document; very dry with a mild condescending tone thrown in for good measure. One interesting thing about the video is that its talking head is actually a man who looks like a woman (sort of). He takes advantage of all those fashion technologies that Kurzweil talked about in order to partially transcend gender, if only superficially. Besides being a transgender transhumanist, ZJemptv is also an atheist. All of the big trends culminate into a singularity here. You know that ideas are powerful when you begin to see them manifest in the lives of real people. That is what makes the ground level tranhumanists so important. They are human barometers, gauging the winds of change with impressive accuracy. We can all thank culture creators like Marilyn Manson and Ramona for paving the Y for these new age rebels.

Rising up from the transhumanist base level, we find ourselves nearing the capstone of their hierarchy. Transhumanist leaders not only know about postgenderism, they choose to embody it. A prime

Postgenderism

example is Martine Rothblatt, who was at one time known as Martin. Martine is a male to female transgender individual, but more importantly, she sits on the board of trustees for the IEET, which is basically a subdivision of the World Transhumanist Association. Rothblatt is described as an "attorney entrepreneur" on the IEET website. She started up Sirius Satellite Radio. Beyond simply starting this company, she was actually a guiding force in producing the legal policies that gave her the ability to do so. She led the charge to establish major international treaties for "space-based navigation services" and "direct-to-person satellite radio." This is no small endeavor, but rather a huge international project that requires the combined effort of separate governments. Rothblatt had the distinction of coordinating this effort, which is impressive to say the least. It takes an enormous amount of influence to get international treaties signed. Rothblatt is a mover and a shaker; moving beyond the confines of this earth to set up operations in space, and no doubt shaking all the right hands on the way. She launched the first vehicle location service in 1983! The private use of satellites to monitor travel was made possible by her work. She also led the International Bar Association's project to develop a draft of the Human Genome Treaty for the United Nations. Around that same time she founded the biotech company United Therapeutics. Rothblatt is a high-level businessperson, high-level transhumanist, and powerful alchemist. Alchemy, after all, can be practical. The world of international business is now literally being carried out in space. Rothblatt is a key player in this enterprise.

Rothblatt's interest in the occult is plain to see if only in the name Sirius. This of course is the Blazing Star that is revered within the Mysteries. It is also known as the Dog Star. The Sirius Satellite Radio Logo is a dog with a star as its eye, a simple graphical play on the name. Beyond this, it is an occult homage and nod to all of those who can *see*. In terms of transhuman occultism, Rothblatt is a founder of the Order[85] of Cosmic Engineers, which is one of the spiritual subgroups within the transhumanist movement. They believe that in the future, pure intelligence will spread throughout space and even create new universes. It will become godlike. This will be an AGI/ hive mind intellect of some kind. It will use "computronium" to bring the cosmos to conscious awareness, and a state of divine oneness. Ultimate singularity will be achieved through ultimate union. This is a high esoteric belief system. Union is a word used to describe the sexual act of regeneration itself. Traditionally, the act of reproduction

---

[85] *Order*, as in fraternal order. This word is of immense importance esoterically.

is carried out through the union of two separate and distinct sexes.[86] Eternal life and godhood is to be had by a being that can regenerate and perpetuate itself eternally. Reducing the regenerative dialectic to a true unity is an occult yearning. This is the esoteric side of postgenderism. It may well be the reason why Martine Rothblatt chooses to embody the union of both male and female within her self.

As a member of the board of trustees of the Institute for Ethics and Emerging Technologies (IEET), Rothblatt obviously has some influence over its operations. It was this institution that published a white paper in March of 2008 entitled *Postgenderism: Beyond the Gender Binary*. Dr. James Hughes and George Dvorsky, two well-known figures within the transhumanist movement, wrote it. Hughes, after all, was the Executive Director of the WTA from 2004 - 2006, and was given the same role at the IEET, which he co-founded in 2005. Like Rothblatt, Hughes and Dvorsky have a spiritual side that is apparent in their many works, which include the "Cyborg Buddha Project," "Trans-spirit list," and more subversively, this very paper on postgenderism.

Appropriately *Postgenderism: Beyond the Gender Binary* mentions the special roles that hermaphrodites, cross-dressers, and neuters have played across the world and throughout history. They have often held positions of spiritual and political significance. This is an extremely significant point. We are seemingly more tolerant today then ever before, but the simple fact is that androgynous "gods" have long ruled over a *divided* population. This is both true in the philosophical and literal sense. Male priests have long worn robes (a female garment), and actual intersexuals have held religious or political leadership roles specifically because of their uncommon gender traits. The idea of androgyny is one of prime importance within the Mysteries. It is the form that many creator gods have taken. Mastery of the art of alchemy is all about realizing godhood in the here and now.

Consider this line from the paper,

> Gender variant people in traditional cultures often modeled themselves on androgynous divinities, or formed special cults to legitimate their gender variance. Male devotees of the Roman cult of Cybele castrated themselves to become a third sex, the Galli.[87]

---

[86] In the human world at least. It is true, hermaphrodites are found in nature. Do transhumanists envy the genetic power of the banana slug?

[87] IEET, *Postgenderism*, page 4

Postgenderism

This is a good example of just how far some initiates went to get closer to god. You must realize that many of these individuals were considered to be enlightened. The bizarre ritual of castration may not make sense to most people, but it is an important part of an esoteric religion; a religion that hopes to rise up like a phoenix in the coming new age. Its practices and rituals are not only to be promoted but also eventually become the norm. A project of mass enlightenment is now underway and postgenderism, oddly enough, is a part of it.

This quote from the paper reveals much,

> Today however, our Enlightenment values and emergent human potentials have come into conflict with the rigid gender binary.[88]

Conflict is a strong word that connotes violence. This word was not chosen arbitrarily and the reason for its use is far more profound than most people would realize if skimming this paper. The violence alluded to here is ancient. It is the male and female relationship itself. Think about duality, dualism, and the very word dual. To challenge someone to a duel is to pick a fight. The very foundation of duality and its most important human symbols, male and female, have been designed as the models for all things wicked from the very beginning. Natural humanity is considered a terrible thing from this point of view. But fear not, now the world is being prepared for a better way, a more perfect union.

It is seriously implied within the IEET document that our male and female specific genders are holding us back, that the traditional male and female life experience is a hindrance to happiness. The postgenderist goal is to use science to transcend the two sexes and create a new posthuman gender. By overcoming natural human genetics, a beautiful new world can be created. This is all part of a new age project of enlightenment. Still don't believe it?

Here is another direct quote,

> Postgenderists contend that dyadic gender roles and sexual dimorphisms are generally to the detriment of individuals and society. Assisted reproduction will make it possible for individuals of any sex to reproduce in any combinations they choose, with or without 'mothers' and 'fathers,' and artificial wombs will make biological wombs unnecessary for reproduction.[89]

Again, this is all about liberation. The liberty to choose a better way to reproduce is not a difficult idea to sell. It certainly sounds

---

[88] IEET, *Postgenderism*, page 2

[89] IEET, *Postgenderism*, page 2

reasonable (there's that wonderful reason again). Reproductive rights have been a hot-button issue for quite some time now, but most people have not seen what is coming around the bend. Progress and change are to be pushed into overdrive with genetic engineering and other scientific advancement, but what is first needed is a good PR campaign to pave the way. Minds must be readied. The transhumanist movement is one small part of this public relations effort and there have been many other movements that preceded it. All of these have been part of a step-by-step, brick-by-brick act of building.

This document describes many radical feminists who had postgender ideals. Their particular viewpoints were, among other things, labeled "cyberfeminism", and "technofeminism." For them, the only way to be completely liberated from the patriarchal system of control was for females to find a way to self-replicate without the help of males. Doing this would require scientific advances that were not commonly imagined in their time.

> The heart of women's oppression is her childbearing and child-rearing roles...To assure the elimination of sexual classes requires the revolt of the underclass (women) and seizure of control of reproduction...so the end goal of the feminist revolution must be unlike that of the first feminist movement, not just the elimination of male privilege but of the sex distinction itself; genital differences between human beings would no longer matter.[90]   -Shulamith Firestone, 1970

Enter the artificial womb and bioengineering. It is these *emerging technologies* which will liberate women from the pains of childbirth. Beyond this, the radical feminists believed that they would also free women from patriarchy itself. Men weren't likely to give up their power over society so the only way for women to get ahead was to scientifically enhance their bodies. This would give them equal or greater power to men. This interesting ambition points to the fact that the yearning for power is truly gender neutral. There are indeed females who seek control over others, but what are the broader implications of this drive? Are the radical feminist pleas for liberation truly liberating or are they the fantasies of control-freak women who are jealous of their male counterparts? It is safe to say that most people aren't interested in creating gender equality at such a literal level. Most are happy with what they have. They are quite happy having male and female relationships but this really isn't about what content people care for. This is all about what a tiny minority desires.

---

[90] IEET, *Postgenderism*, page 9

Postgenderism

The Bohemian mindset has been let loose on us all and consequently many average people now find themselves becoming its greatest advocates. They too are beginning to feel discontented with the status quo. This isn't a problem itself, the true problem lies in the direction that this general feeling of unease is directed. All special groups that seemingly pop up out of nowhere provide direction for this angst. A healthy distaste for life out of balance is thus twisted and used as a weapon by the Master Alchemists.

Biological gender and "heterosexual desire" are considered a burden to some. It is exactly these things that Judith Butler wanted feminism to do away with. The IEET paper points this out, and agrees with her in the sense that transcending gender will liberate us,

> but only the blurring and erosion of biological sex, of the gendering of the brain, and of binary social roles by emerging technologies will enable individuals to access all human potentials and experiences regardless of their born sex or assumed gender.[91]

It sounds so wonderfully progressive, doesn't it? To practically achieve these ideals will require hard work. More appropriately, Great Work.

As seen in the above quotation, the literal reconstruction of male and female brains is the idea. Male and female specific traits are "coded in the brain," and this is why neuroscience will have to be used to reach true postgender liberation. Remember the profound significance of mind over matter? If someone is biologically and neurologically wired to be a male, then they will act accordingly. However, if there were a way to alter hormones, and actual brain function, then a man could be made more feminine. This is the idea. To totally break down what it means to be male or female. The postgenderist argument is that this will clear up all of the terrible inequality that has existed for so long between men and women. They will finally be able to understand and relate to each other through the

---

[91] IEET, *Postgenderism*, page 3

Fig. 8.2 This illustration is from Michael Maier's *Atalanta Fugiens*, an alchemical classic. Here we see the two headed androgynous man burning up the night.

use of transhuman enhancement. The terrible violence that has long existed between them will be technically done away with. The typical sales point here is that it isn't fair that women aren't included in the workforce as readily as men are. The radical postgender/transhuman solution to this *problem* is to completely redesign our biology so that male and female blend seamlessly together.

What is truly disturbing about this proposed postgender project is its relationship with actual sociological trends. Both male and female roles are now more complicated and confused than ever before. The IEET paper is keen to point this out. It mentions the fact that police, government, and weapons have reduced the need for the typically strong and aggressive male. The strength of technology itself is actually doing away with the need for the old man. According to Hughes and Dvorsky, this social change is good and, "is now being complemented and completed by technological means." The big secret here is that this is all part of one large body of work. The completion of the alchemical Great Work is exactly what they are

Postgenderism

mentioning (although it is very possible that they are not fully aware of this themselves). The hard fact that most people cannot come to terms with is that our very lives are the subject of a massive science project. Our world is the Master Alchemist's crucible, in it we are being mashed up and put through the various stages of the alchemical process to ultimately become gold. We are being putrefied and sublimated. Our male and female genders are to be blended, both philosophically and physically. The protective role of the man has been reduced on purpose.

It is the job of men to protect their tribe. This crucial aspect of human life has been socially reduced for a reason. We are slowly being led into a world in which everything is to be provided by government. We are steadily becoming completely dependent on this artificial system. If the technoprogressive project continues further, then we will eventually end up in a complete nanny state. This is the truth behind the egalitarian world utopia we have heard so much about from politicians, academics, and entertainment. Both male and female strength have been targeted for destruction. Not only have our gender roles been confused, we have actually been chemically bombarded as well. Chemical warfare has been waged on us to destroy our biological gender. Yes, the IEET paper makes mention of this chemical attack, but it doesn't mention that it has been done purposefully,

> The incidence of intersexuality is disputed, and it may now be more prevalent than before due to environmental chemicals that mimic estrogen and interfere with fetal genital development,[92]

Again, that was Hughes and Dvorsky in their *Postgenderism* paper. In an online text interview TransAlchemy conducted with Dr. Hughes, he was asked what chemicals were being referred to here. Hughes said they were,

> pseudo-estrogens and endocrine disruptors, specifically phthalates from plastics.[93]

Now why exactly are these plastics all around us? BPA/phthalates have been present in food and water packaging for a very long time. We are given to believe that this has all been some big mistake, something that may be fixed. We aren't supposed to question why these toxins have been used so widely in the first place. Different

---

[92] IEET, *Postgenderism*, page 3

[93] http://www.transalchemy.com/2009/07/transalchemy-intervies-drj-hugeson.html

sorts of plastics could have been used instead of those containing BPA and yet they weren't. Our constantly lowering sperm counts and fertility levels are supposed to seem coincidental. Beyond this, our entire existence is supposed to be viewed as coincidence. This is the concept that science has drummed into our brains. We aren't supposed see any rhyme or reason to our own life, let alone of anything else. To people stuck in this reduced state of consciousness, the huge media promotion of cosmetic and gender reassignment surgery seems like just another coincidence. They can't entertain the notion that things happen for a reason.

More evidence that chemicals are being used to postgender ends is mentioned in another quote from the IEET paper,

> Efforts to treat female depression and male aggression, autism and A.D.D. would give us ways to make the brain more androgynous.[94]

Of course *us* is supposed to be perceived by the reader in its democratic sense. We are supposed to think that postgender transformation will be a collective free-will decision made out of our yearning for equality. From this point of view, it really is us (you and I) who will make this decision. Another, and far more realistic, way to read *us* is in the context of an idealistic bunch of technocrats who have taken it upon themselves to improve the human race. Do Hughes and Dvorsky use the word *us* believing they are a part of this group? We can't say for sure, but their views on this subject are troubling to say the least.

Chemicals are sold to us under the guise of improvement. Prozac has become a quick and easy fix for various mental problems. A dangerous trend has been established in our society: to eradicate any and all pain. Mental disorders such as depression are explained one-dimensionally as ailments. They are viewed as randomly or genetically occurring diseases, and not as symptoms of a larger sickness. Maybe, just maybe, such discomfort is an indicator that something is wrong with our environment. The prevailing method for medical practitioners today involves labeling the patient as the ultimate problem and to completely ignore the true causes of various diseases. Transhumanism follows a similar model that appeals to our feelings of self-doubt that model being: you are broken, but we can fix you. Psychopharmacology (drugs used for psychological disorders) is certainly multi-faceted. The postgender angle is something that even the most ardent critics of the medical industrial complex could easily miss. It comes as no surprise that Ritalin

---

[94] IEET, *Postgenderism*, page 12

pacifies young boys and makes them less of a threat in the classroom, but the long-term effects of this and other such drugs are not always understood. Ultimately, these boys will become men, but what kind of men will they develop into? They are being pacified throughout their lives and on into adulthood so that they will be disabled. They will be less able to stand up and protect their tribe. This leaves everyone vulnerable. We have now been disabled to such an extent that the existence of natural enemies and all hostility goes completely unnoticed. We never even see that we are in danger. The very idea that we could be is often condemned by popular opinion. This is a testament to the mess that we are in. Men and women have let their guard down for the sake of gaining petty rewards in this faith-based system. We have gone along to get along. We have taken from the hand of the greatest deceivers, the deadliest predators, and never even realized it.

It cannot be stressed enough that transhumanism/postgenderism (and many other isms) are all part of one large plan. We are being made to hate ourselves. Our actual human strengths are being explained away as weaknesses so that we believe a great lie. That great lie is that we are indeed weak. This deception will make us crave false strength. We will truly want technological enhancement. The sad reality is that this so-called enhancement will simply enhance our ability to do what we are already doing, and that is efficiently serve our dark Master which uses and abuses our true value. Instead of questioning his faults, we are instead turned around and made to question our own. Many of our supposed faults are merely the products of Madison Avenue. Wicked genius has sold us our depression, the very same thing that the medical establishment tells us is our own genetic fault. There is purpose in this entire process! We are never supposed to look at the true light of our glowing society. We are warned that this would blind us. Actually, if we did turn our heads toward this light, all we would see is darkness. A dark void of self-loathing propaganda would stand before us. We would understand that psychological warfare has been waged upon us all, that we must fight to preserve our most basic human rights, and in fact our very human identity. The natural world, of which we are all a part, is being done away with in favor of complete artifice. This dark masterpiece is the work of enlightened visionaries.

There is no shortage of genius within this campaign of liberation. It wisely plays to our yearning for equality. The radical feminist criticisms of our patriarchal system are not completely unfounded. We are indeed living in an unbalanced power structure. Men have been leading the way for a very long time (thousands of years). Instead of having both men and women in leadership roles, we have been overrun by monopoly men. The aspects of human life that women are more naturally inclined to take leadership in have been taken over by men. In our ancient past religious and spiritual roles were filled by women. It is these roles that the monopoly men commandeered. Long ago they removed women from the human power dynamic. They donned female garb (robes), and began preaching the "word of god." God came to be known as the *Father*. Over time, this became completely normalized. Throughout the world, the idea of a holy man is common, but holy women are not as prevalent. This is evidence of a profound problem.

What we need to realize is that men and women are indeed different, but this is not a bad thing. We have been trained to associate mere differences with inequality. Men and women are now supposed to be able to do all things equally. This means that we are all supposed to fall in line with a completely unified status quo: go to work, make money, and earn our own way. This ridiculous formula is a perversion of both male and female roles. It is an unnatural state for us all. Our natural roles have been stolen from us and in their place a completely artificial form of life has been crafted.

Women hold the sacred distinction of being the bearers of our children. For obvious reasons this has always been a profoundly important aspect of human life. It is the female gender that has always held this great power and all of the secondary powers that come with it. Could it be that control-freak men have long been greedy of the biological power that could not be taken away from women? Females are literally closer to children because it is within in their bodies that children develop. Innate nurturing abilities are reflected in a number of ways. Of course, women are the ones who take primary care of children but beyond this, they also take care of the entire tribe. They are more emotionally and spiritually inclined than men as well as being more intuitive. Because of this, they have traditionally been given sacred roles, which have protected the human family in a uniquely feminine way. They are our lost spiritual leaders. It is because of an unbalanced male dominated system that most people now view any sort of spiritual concepts as ludicrous. Over a long period of time, male dominated spirituality has led to its ultimate conclusion. Believe it or not, this is what we are seeing come to a head today. You can't put a bunch of men in charge of everything and expect spiritual balance. You will inherently tip the scales and end up with men who substantiate their spiritual power by actually taking on feminine traits. This is the perfected/androgynous man of the western occult tradition, which is a self-perpetuating model of male dominance. One real life illustration of this phenomenon is seen at the all-male summer camp for the elite, Bohemian Grove. One of the many examples of their phallic (self) worship is seen in their staged plays. Due to a complete absence of women at the Grove (even amongst the staff), the men have to use their imaginations a bit. During these rituals to Dionysus the men play both male and female roles. This shows what power-grabbing men have done for thousands of years. Women have been pushed out of the power equation. During the Grove's festivities, there are frequent derogatory comments made about real women.[95] Is it any surprise that male prostitutes are brought to the party in bulk? No girls allowed! It all seems so juvenile but when you realize who these men are, and what they do, a whole new level of reality sets in. These are the men who run the world. What exactly are they up to?

It is no secret that women have been marginalized in this world, but the big picture remains elusive to most people. The very foundations of our entire system are scantly known. This is why the true oppression of females has been hidden. The radical leaders working within the Women's Liberation movement have many times turned out to be carbon copies of power hungry males. Their message has been one of equality but again, this equality is merely uniformity. The idea is to make men and women the same. The Judith Butlers of the world have wanted women to become men. They are ironically acting in a masculine way, reaching for the perverted power that this system has well established already. They have been driven by the same greed that drives the men of Bohemian Grove. They want a female-dominated society, but this is no solution at all; it is merely a reflection of the real problem. Their matriarchal paradise would be just as unbalanced as this patriarchy. Their message has actually been geared toward destroying true feminine strength. The abolition of natural birth was the end goal they sought. In opposition to these radical factions have been feminists of another view. They understand that the female's ability to give birth is extremely important. If given up, our male dominated system could actually be taken to its absolute limits. Women would become physically obsolete. Don't think for a second that this idea does not excite the monopoly men. They drool over it. The ability to create life by themselves would enable them to destroy women once and for all. This actually represents the scientific

---

[95] Watch the Alex Jones film *Dark Secrets Inside Bohemian Grove* to see this for yourself.

completion of an ancient work and it has been happening right under our noses. The "Will of God," also known as the father, is overcoming *Mother* Nature. The artificial womb is coming. It has in fact been in the works for a very long time. Radicalized feminists were but one faction of shock troops sent out to pave the way for its acceptance.

The Artificial Womb

J.B.S. Haldane coined the term *ectogenesis* in 1924, a word used to describe the birthing of humans with artificial wombs. The most important benefits of this future technology were laid out from its very first mention. It was Haldane who saw ectogenesis as a necessity for combating the high fertility levels of the "less desirable members of the population." It should come as no surprise that ectogenesis was and still is a eugenic project. By controlling genetics, all stupidity was to be eradicated scientifically. Stupidity was a word used to describe the bulk of the population. By genetically engineering and generating better humans with artificial wombs, the "inferior types" would be done away with by default. It has long been the unwashed masses that were to be cancelled out so that a new age utopia could be brought in. Remodeling human generation through the use of artificial wombs and reducing the amount of natural human births was the way to make this scientific utopia a reality.

Just five years after Haldane let the ectogenesis genie out of its bottle, a very interesting article appeared in Cosmopolitan Magazine written by the Earl of Birkenhead,[96] an English aristocrat. The article was labeled in this way,

> It is a Forecast *of* What This World Will Be 100 Years From NOW

It actually mentioned genetic engineering, ectogenesis, and many other scientific achievements. Like Bacon's New Atlantis, this piece clearly described technologies that had not yet been invented in the year 1929. Birkenhead describes television, and even calls it by that name. He proclaims that everyone will have one by 2029. The most interesting part of the article has to do with the artificial womb and what it will be used for,

> It is possible, however, that by 2029 the whole question of human heredity and eugenics will be swallowed up by the prospect of ectogenetic birth.

---

[96] *Save This For Your Children's Children*, Cosmopolitan Magazine, Feb. 1929

In other words, genetic engineering would be perfected alongside the artificial womb.

> By intelligent combination of suitable genes it will be possible to predict with reasonable certainty that truly brilliant children shall be born of a marriage,

This is what eugenics is all about, perfecting the act of regeneration and seeing that all inferiors are disposed of intelligently. What better way to do so than through scientific method and technology? This very idea is being marketed to us today as a completely humane way of eliminating suffering, stupidity, and all the rest of our bad traits. We are being sold on eugenics.

The Earl of Birkenhead knew that artificial wombs would come but he also knew they would be opposed,

> The possibility of ectogenetic children will naturally arouse the fiercest antagonism. Religious bodies of many different creeds will rally their adherents to fight such a fundamental biological invention,

This very concept is well known today within the research community that is actually building artificial wombs. Research has been purposely kept quiet so that people will not protest it en masse. The research community is waiting for the proper time to unveil the fruits of its labor, but more on this point later.

His Lordship continues to say of ectogenesis,

> it will separate reproduction from marriage, and the latter institution will become wholly changed.

There are plenty of details on this point in the IEET paper. The reworking of the *social contract* is a huge aspect of all this. The legal definitions of everything, especially intimate human relationships, are to change. Society itself is to be totally reengineered. Human biology is to be tailor made so that it fits the measurements of the Great Architect. It is on this point that Lord Birkenhead really lets it all hang out,

> Further, the character of the future inhabitants of any state could be determined by the government which happened temporarily to enjoy power. By regulating the choice of the ectogenetic parents of the next generation, the Cabinet of the future could breed a nation of industrious dullards, or leaven the population with fifty thousand charmingly irresponsible mural painters,

He goes on with his eugenic predictions by saying,

> A further immediate consequence of ectogenesis would be a plea that society should be allowed to produce the human types it most needs,

instead of being forced to absorb all the unsuitable types which happen to be born.

This is a recipe for complete and total control, a calculated, scientific and mathematic formula for sustaining society. There is nothing spontaneous or intuitive about it. It represents the perfected male-dominated world, where everything must be completely measured. Birkenhead asks,

> If it were possible to breed a race of strong healthy creatures, intelligent to perform intricate drudgery yet lacking all ambition, what ruling class would resist the temptation?

This is indeed an excellent question. Here you have an actual member of the "ruling class" telling you that the creation of a scientifically designed slave is simply irresistible. You had better question the nature of human enhancement now, while you still have the capacity for any kind of "ambition." Birkenhead said,

> Every impulse which makes slavery degrading and irksome to ordinary humanity would be removed from his mental equipment. His only happiness would be in his task; he would be the exact human counterpart of the worker bee.

The animal symbol of the bee is used frequently to describe the perfect laborer. It pops up again and again within eugenic writings because it is an ancient symbol of the perfect citizen and the beehive the perfect civilization. A perfected society would buzz away. All those tasked with mundane labor would be tuned for performance like machines. A large portion of this article is an ode to this "ectogenetic Robot," but at the end Birkenhead explains that such a machine may never come to bee. Instead of human robots,

> It is far more likely that men will work as machine-minders for one or two hours a day and be free to devote the rest of their energies to whatever form of activity they enjoy.

Hooray, the transhuman paradise we're all hearing so much about these days! Be sure to realize that his Lordship is certain that children will be produced in laboratories. Genetic engineering will perfect the eugenic process of biological cleanliness.

The mental conception of a new method of physical conception came about in the early twentieth century (at the latest). There has continued to be talk of the artificial womb all the way up to our current day. In the 1950s and 60s a scientific journal entitled *Utopian Studies* debated this topic. The very title of this journal speaks volumes. This is all part of a utopian project which is known by many names. Its work is divided into many categories. A more perfect singularity has been the goal. This has long been known but the term singularity is only now being released for mass consumption. The reason for this is that we are going through the final motions of a plan. A scientific age of reason is upon us (if we wish to see it through). Understand that there is direct purpose behind all of this. The very word *reason* implies this.

> The artificial womb represents the completion of an even longer historic process that began nearly 400 years ago at the dawn of the scientific age. It was Francis Bacon, the father of modern science, who referred to nature as 'a common harlot.' He urged future generations to 'tame, squeeze, mould and shape' her so that 'man could become her master and the undisputed sovereign of the physical world.' No doubt some will see the artificial womb as the final triumph of modern science. Others, the ultimate human folly,
>
> -Jeremy Rifkin, *The End of Pregnancy*, London Guardian, Jan. 2002

This piece by Jeremy Rifkin laid out many facts behind the artificial womb research of Cornell University and Juntendou University (Japan). This research was appropriately compared to the writings of Sir Francis Bacon. Rifkin is the President of the Foundation on Economic Trends in Washington D.C. He has written books about this subject himself and is well aware of what is happening. In this same article he predicts that we are likely to see widespread use of artificial wombs within a generation. Most women would now be opposed to such things, "but their expectations might represent the dying sensibilities of the old order," Rifkin writes.

That's a charged sentiment if ever there was one. The destruction of the "old order" is a common theme regarding this subject matter and of course, who hasn't heard of the New World Order? At this point the idea of a New World Order feels old and even trite, but it must be understood that the very idea has been unleashed on us all with definite purpose. A New World Order is emerging and our psychological compliance is being manufactured slowly but surely through simple repetition. It is expected that we will eventually break down and accept this entire New World Order project by the time that its actual infrastructure is in place. Ectogenesis, transhumanism, and postgenderism are all aspects of this Brave New World Order. At the time it was published, Huxley's fiction was an absolute horror story. A few decades from now it could be praised as a beautiful, prophetic, and lovely work that was ahead of its time.

Some serious corroborating evidence for all of this can be seen in a section of the fall 2008 *Harvard Science Review*. Under the heading

Focus: Brave New World is an article entitled *Artificial Wombs - Delivering on Fertile Promises*. This title is obviously a direct reference to Aldous Huxley's *Brave New World*. Our real world is already following Huxley's fictional model as this article explains. "Delivering" (i.e. delivering a baby) "on Fertile Promises" (which are centuries old) are the actual scientists who are now only beginning to let their quiet research projects out of the bag. The general public is slowly being made more tolerant of certain ideas.

Consider this quote from the Harvard article,

> So why has relatively little attention been paid to research that could lead to a major change in reproductive capacity? In fact, much work goes unpublished because of the uproar it might create among activists, politicians, and religious figures for its social implications.

There you have it. How disturbing is it that this project has purposely been going on in the dark so that the public would not be aware of it? What happened to democracy? How will a supposed technoprogressive democracy be built upon a secretive foundation? The vehement opposition to artificial wombs spoken of here is the exact same thing that Lord Birkenhead mentioned in 1929. It has long been known that people will be strongly opposed to artificial wombs. The solution to this problem has been to do the work privately, in the dark, and away from the prying eyes of the public. However, those following this method have not been afraid of using public funds. After all, tax dollars are what fund most research universities. Even "private" schools such as Cornell thrive on big grant money from the NSF et al. It seems clear that we are paying for crypto-eugenic research whether we like it or not. This again points to the fact that democracy is the biggest joke on the face of this planet. It is pure lip service, conjured up to give us a false sense of security. A scientific technocracy is slowly rising to complete power. Its worker bees up to this point have been a batch of morally ambivalent scientists carrying out their labors for the sake of progress. To be blunt, I don't think many of them really care about the broad implications of their personal labor. It just isn't something that they ever consider. On the other hand, there certainly are scientific laborers who have been convinced that their work serves a greater good. Specialization makes it difficult for many of these well-meaning scientists to see the big picture. They have been unfortunately blinded by their Ivy League-induced tunnel vision. Most have become slaves to technique itself, and in the process have abandoned something very important. It is morality, which has been marginalized, in this scientific system. Commitment to progress is required to get the job done and so morals must always take a back seat to tangible results. Besides this, morals have been explained away as something totally relative and therefore not worthy of attention. Scientific ethics are questionable.

This article, which comes directly from Harvard, is a glaring example of this point. It coldly lays out logical methods for creating and implementing artificial wombs. It comes complete with pictorial evidence of the same research that Rifkin spoke of in his London Guardian piece. The image of a premature goat lying in an artificial placenta is seen on page 36. This was the work of Professor Yoshinori Kuwabara of Jutendo University. It is a testament to the fact that the artificial womb has remained a hidden reality. The actual construction of this technology is accompanied by the usual rationalizations for its use. Colleen Carlston, the article's writer, is just full of great reasons to use artificial wombs,

> Beyond imagining new world orders, there are also some pragmatic reasons for why artificial wombs might be employed.
>
> -Colleen Carlston, *Harvard Science Review*, Fall 2008

There's that wonderful New World Order again. The justifications found here are nearly identical to those found in the IEET paper on postgenderism: artificial wombs will be "safer" for mothers, they won't have to leave their careers, and of course a host of other eugenic corrections on reckless motherhood follow. In the name of safety, artificial wombs will be used to ensure that bad mothers don't taint the fetal development of their children by consuming drugs or alcohol. This added safety will tie into the medical insurance industry. Insurance companies will promote the use of artificial wombs by giving benefits to those who opt to use them. This is the wonderfully safe and secure New World that we are stepping into. In all aspects of life, security is becoming the selling point for a new system. Just who has the capacity to offer total security to a concerned population? A public/private partnership between government and corporations is already here and its practitioners are more than willing to raise any children that we are too busy to care for ourselves. They will keep them safe. This is the *nanny* state. It is the new family structure for the New World Order.

One of the many overtly eugenic disclosures in the Harvard Science Review is the notion that the artificial womb may actually be too good. It could be used as an alternative to abortion and thus "overburden" adoption agencies with too many children. This article cites a quote made by Helen Pearson from her work *Making babies: the next 30 years* in which she says,

> There are around 1 million abortions per year in the United States and there would have to be labs throughout the country, but if we put all these in artificial wombs and then put them up for adoption we would have one million more babies. It would be a nightmare

A nightmare; what a strong word. It is extremely telling that in the minds of those who most anticipate the arrival of the artificial womb, increased fertility is a horrible thing to be avoided. This is the life/death dialectic of eugenics. There must be life to the strong and death to the weak. The Malthusian population catastrophe can be seen here as well. To avert disaster and institute security, we have to accept a healthy dose of death. We must all sacrifice for the good of the race. The people working on and promoting these technologies are eugenicists. We should all realize that the artificial womb and the declining fertility levels of Western nations go hand in hand. Our population numbers have been declining for quite some time now but this is not good enough for the hardcore eugenic cult leaders. They crave more.

The Harvard article also mentions a quote from Regine Sitruk-Ware, a reproductive endocrinologist at the Population Council in New York,

> If we look at centers in reproductive sciences funded by the National Institute of Health, there are more than twenty on I.V.F. and only a handful on contraceptive research. It's more politically correct to help people get babies than the reverse, but it's important to have a balance.[97]

Now just what is going on amongst the ranks of the Population Council? Where did this council come from? You might want to look into these things. Whoever they are, they apparently hope to strike a "balance" between the eugenic life/death dialectic. This ties in directly to the creation of the artificial womb. The field of genetics began with a eugenic bias. The eugenic philosophy has sustained itself to this very day and can be seen throughout the biological sciences. It is this philosophy which is slowly creeping into the minds of an educated public. The truth of this is self-evident, however the devil is in the details, and fancy terms like transhumanism have fooled most people into believing that eugenics is dead and gone. No, it lives on, if not in your mind, than in the mind of your child. The progressive education system has seen to that.

---

[97] Harvard Science Review, Fall, 2008, page 39

The Future of Sex

Throughout the ages the male and female relationship has been a simple fact of life. It is one of the natural foundations upon which our entire existence is based. All legal institutions that have been built up around it are nothing compared to the actual bond itself, the joining of the two sexes. We are now beginning to hear arguments from some who claim that the biological roots of this ancient bond are not as sacred as we thought they were. The idea of completely *trans*cending our male and female genders is being promoted. This idea is coming from individuals who seek a more perfect union of man and woman through chemical marriage. They wish to chemically bond both man and woman so that a singular new creation is born. This would be the scientific achievement of an occult ritual. Gender singularity is now within the realm of possibility. Do you have any idea how monumental this is? Merely classifying postgenderism as a progressive or liberal movement would be downplaying its significance extremely. It is a part of an ancient esoteric religion.

Like all other aspects of the transhuman project, postgenderism is about the attainment of godhood. Creator gods throughout time have been androgynous and now there are men intent on achieving that very state. They want to be able to "make love out of nothing at all," like only a god could.

We have to understand the magnitude of change that has happened within our own time. Ideas that would have incited vicious contempt from most people decades ago are now a part of common conversation. Serious subjects are met with a shrug and a "meh." Today's youth have been bombarded with so much shocking and sexually explicit stimuli that they now suffer from a very serious condition. A state of complete indifference has been achieved through sensory overload. Chaos and confusion have been unleashed upon the world to overwhelm us. Naturally, many have become *confused* sexually. Most have lost their reverence for sexuality. Its divine nature has been debased and correspondingly, we have all been degraded. We have steadily lost touch with our natural human identity. This has caused the so-called old order to crack and crumble. The question now is: do we want to continue this project of destruction? Do we want to break ties with all that we are in order to transform into something better?

The sacred act of regeneration is now being recreated. This is in fact part of a high alchemical ritual. Over time our actual thought process has been transmuted. It is humbling to see that our attitudes toward sex have undergone rapid change within a short amount of time, a span of decades. As the Singularity approaches, this effect begins to speed up.

Just think about 1950s America, and its attitudes toward sex. During this time a traditional and very conservative view was contrasted by a steady erosion of old taboos among the youth. For the most part, the older generation taught their children about sex in a very strict manner. The sex act itself was coupled with a set of strict guidelines. No one was to do it without first going through the legal process of marriage. The institution of marriage itself must be understood as nothing more than a social contract/legal (and ultimately fictitious) overlay that has been hoisted upon a divine bond. Anyone deviating from the social norm of the day was to feel great shame and be shunned by civilization. The imposition of these regulations upon the youth caused great conflict between the generations as the youth were meant to feel ashamed of a completely natural process. These harsh rules fueled the fires of rebellion within young hearts. Young adults sought out ways of liberating themselves from this frustrating situation. Liberation came conveniently prepackaged in the form of popular culture. The old cliché of rock and roll's corruptive influence is certainly true, if only partially. The sexual revolution of the 1960s took the rigid value system of the past generation and turned it on its head. "Free love" was seen as an escape from the harsh rules of the old value system. Pent-up sexual energy found a release in the flower-powered orgy of the hippy movement. The general idea of feeling good was promoted with the proliferation of recreational drugs. This represented a major change in attitude as we collectively walked through the *Doors of Perception*. Many believed that this was truly liberating. They did not fully understand what was behind the Door, even after opening it.

As the decades have passed, our ideas of sex have continued along a steady path of change. The baby boomer generation is now old and their liberal sensibilities toward sex have allowed a relaxed laissez-faire attitude toward further sexual revolution. Sex itself is increasingly associated with pure recreation (not creation). It has become a form of mere entertainment. The ecstasy of sexual intercourse has been perverted and taken as an end in and of itself. The utilitarian aspect is, of course, still there, but perhaps not for much longer. Overall, the identity of sex has been altered significantly. It has become a casual event. The concept of "casual sex" was explicitly promoted in the 1990s. This propaganda

Postgenderism

campaign was seen in a barrage of popular culture that depicted the seemingly lighter side of sex. Think of your favorite 1990s sitcoms. Nearly every one of those programs deployed a rapid-fire assault of one-liners and humorous sexual situations. As casual sex was casually depicted time and again, procreation became an afterthought, a joke in its own right; what this means to the human race as a whole should not be understated.

It is completely accurate to say that the road we are winding down leads to somewhere very strange and disturbing. We are now in serious trouble. Transhumanists would have us believe that simply having this opinion represents a regression to old and outdated values that no longer suit the good of our species. In other words, not only are we to be completely tolerant of all this, we are to see it as beneficial. Understand that you have every right to be upset about these things and that you can hold any opinion you want. The fact that people would never have accepted any of this one hundred years ago is not because they were primitive savages, or unenlightened prudes, it is because they still had some semblance of a survival instinct left within them. Political correctness (and chemical warfare) is disengaging this instinct. The wolves in sheep's clothing have been sent out to pacify us all. Psychological warfare has been waged and we don't even realize it. We have actually been psychologically disarmed to such an extent that we aren't merely lying down to die, we are opting to do the job ourselves by committing suicide.

If only we could view this practically, like good architects, then everything would fall right into place. Virtual reality could quickly and easily pick up where our old lives left off. This would be better, safer, and more diverse in every way imaginable. Why would we not choose this? Well, let's think about that.

We have to understand how far the virtual world has developed already, and where it will go if left unchecked. The end goal (for the bulk of the population) in terms of postgenderism has much to do with virtual reality. As Hughes and Dvorsky tell us,

> Trans- or post-humans would at least be able to transcend the limitations of biological sex, and would eventually be able to transcend the biological altogether into cybernetic or virtual form.[98]

The virtual realm is where all technological paths converge. At this point, it would be simple to list the usual benefits of technological enhancement that transhumanists are always talking

---

[98] IEET, *Postgenderism*, page 7

about, but what is far more interesting is the specific promotion of virtual sex that is found in the IEET document.

The paper describes in detail how virtual sex has already softened us up, but it obviously isn't described as such. The typical positive spin given on electronic sex is that it is safer, easier, more convenient, and completely tailored to individual desires. What isn't mentioned are the sad consequences of alienation, perversion and addiction that come along with this "safe sex." Virtual sex is so neat and tidy that it may sanitize the great mess that biological humanity has made of this world. Across transhuman literature this theme can be found. Leading transhumanist and AGI researcher Ben Goertzel put it this way in his *Cosmist Manifesto*,

> It may seem cold or eccentric to say so, but the fact is that orgasms and genitals and romantic relationships - as glorious as they are - are hideously badly designed... It's obvious that other entities serving the same purposes (and more) could be developed.

Old world sex is actually being depicted as dirty and outdated. However, this doesn't mean that it can't be exploited in virtual realms. Any and all terrible perversions can be cleaned up there, not only to make people happy, but also to keep them safe. Tada! No more sexually transmitted diseases or unwanted pregnancies. A very important aside lies in the fact that pregnancy and disease have been associated with one another for so long now (being listed side by side on lists of sexual risks) that most people are actually beginning to view them as the same thing. In other words, pregnancy is now a disease. The opinions of mindless bloggers and youtube atheists can corroborate the fact that this meme is well established. Understand the powerful process of association that constantly occurs within our minds. It is through simple and repeated associations that we change our opinions and our behavior. This fact of human psychology is extremely far-reaching. Simple subconscious association is at the root of all perversion.

What happens when sexual desires can be scientifically catered to with incredible efficiency? This is not a simple question to answer. A great many consequences emerge. The destruction of actual human contact is one outcome of favoring the virtual over the real. It is exactly this sort of alienation that has been fostered in our real world. We have been steadily drifting toward complete virtualization of sex (and many other things) for a very long time. Remember that door which was opened back in the 60s? That was actually a high occult ritual which has brought us to where we are now. It altered our perceptions thoroughly. Free love has steadily morphed into complete depravity. Thanks can go out to Hugh Heffner and all those other wonderful purveyors of pornography for their contributions to the Great Work. They have paved the way to bigger and better things.

What do we think will happen when any sexual desire can be accessed instantly? It is not hard to understand how addictive and perverse this cruel joke can become. The exploitation of human sexuality knows no bounds and the IEET document gives chilling reference to where we may find ourselves in the near future,

> The growing sophistication of AI and robotics to detect human emotion, anticipate human desires and respond in ways that simulate a human response will also speed the virtualization of sex.[99]

It goes on to cite work concentrated on the perfection of virtual sex devices. What do you think will become of all this?

While we have been busy fulfilling our wildest desires in virtual reality, some very serious work has been done in actual reality. The complete destruction of the human condition can only be achieved by removing its defense mechanisms. Men are designed to protect their tribe. This protective instinct is incorrectly labeled by postgenderists as meaningless aggression. It is not meaningless; it is an integral part of the survival of the species. They also grossly misrepresent women's natural inclination to be nurturers. Taking care of a child is portrayed as a hindrance when in reality it is a beautiful and life-sustaining thing. Postgenderists and transhumanists sincerely wish to see the complete and total destruction of the male/female relationship and the institution of a more safe and secure nanny state. They don't mind that our natural ability to bond with one another has been broken. Family life, as we know it, is being eradicated. A high-tech war has been waged on humanity and now is the time to start calling it for what it is. The truth is right in front of us written in plain English,

> But the final liberation from dyadic, gendered, heteronormative relationships will likely come about through use of drugs that suppress pair-bonding impulses,

> -George Dvorsky, Dr. James Hughes, *Postgenderism: Beyond the Gender Binary*, Mar. 2008

---

[99] IEET, *Postgenderism*, page 11

What kind of liberation is this?

-Chapter Nine-
A Worldwide Death and Rebirth Ritual

The most basic of human relationships are now being put through the grinder so that all other natural social groups will be destroyed by default. If you target the most basic foundations of life, then the consequences of your attack will be far-reaching. Before creating something new, you must first destroy the old. This is the method by which artificial creation works. Destruction itself is symbolized in the dissolution and putrefaction stages of alchemy.

In a book about alchemy, published by Taschen, the "Opus Magnum" is discussed. The "Great Work" is explained as being all about combining the two opposing forces of "materia prima" into a perfect union. What is being discussed here is the actual male and female relationship. It is this fundamental human bond, which is eventually to be broken, so that it may be welded back together according to the design of master craftsmen.[100] An androgynous god would thus emerge from the destruction of the flawed duality of man and woman. On the very same page of this book, a quote from 1778 by Büchlein vom Stein der Weissen reads,

> First we bring together, then we putrefy, we break down what has been putrefied, we purify the divided, we unite the purified and harden it. In this way is One made from man and woman.[101]

Did you get all that? In case you missed the highly charged symbolism, look again. Be sure to read between the lines...

Putrefy, purify, divide, unite, and *harden*: these are all alchemical terms, and their meanings are highly symbolic. Think of this entire quote in terms of marriage itself. Understand that what is more important than actual marriage is what preceded the idea of marriage, and that was the natural relationship between man and woman. It was exactly this foundation of human life that the Master Alchemists were intent on altering. It is the alchemists who brought man and woman together in a way, which suited their own ends, with the legal contract of marriage. It must be understood that our modern concept of

---

[100] Craftsmen, builders, or architects.

[101] Roob, *Alchemy and Mysticism*, page 111

marriage is merely a legal contract. It was the alchemists who wrote up this contract. Once marriage was designed according to their legal specifications, it was easier to directly affect the relations of men and women. It was possible to aggravate both men and women through various social contracts, expectations, and other pressures. We have been witnessing the breakdown of the institution[102] of marriage for many decades now, and this is not by accident. All of the little everyday problems of the modern world add up and make the maintenance of any relationship extremely difficult. This pressure corresponds to the putrefaction stage of alchemy. Putrefaction is when things go bad, when they stink and decay. To "break down what has been putrefied" is to breakdown the psyche of the separated man and woman, now left to their own devices in a cold world. During this stage, people lose hope. People who are indoctrinated to actually believe that marriage (a mere legal contract) itself is holy, upon putrefaction are forced to make sense of their situation somehow. Few are able to see that the institution of marriage was manipulated so that man and woman would actually separate from one another, and so they blame themselves. To "purify the divided," in one way is seen in the example of giving women hope in the false security of chasing a career. The proclaimed equality of men and women actually equates to the destruction of their unique and important gender roles in favor of artificial replacements. The ultimate completion of the work comes with the union of the two opposites, male and female, who at this point have been separated for quite some time and are involved in living their own lives. The individual's "purity" is the role that the system itself gives to each unique man or woman. The creation of a postgender posthuman would be the literal way to "unite the purified and harden it." The hardening is the setting of the postgender ideal, the literal attainment of the androgynous New Man.

Sounds bizarre doesn't it? Well it is, but the strangest thing is that this truly is the direction that we are all being led collectively. It is a strange and ancient ideal that we are being made to fit. This is a high occult labor, a work of ages. The alchemists believe that androgyny is a spiritual state of perfection, something for a god to enjoy. To them, its literal attainment is a way to become One with God. It represents ultimate Singularity.

> That man was primarily androgynous is quite universally conceded and it is a reasonable presumption that he will ultimately regain this bisexual state. –Manly P. Hall, *The Secret Teachings of All Ages*, 1928

---

[102] Note the very word *institution*, what it means, and that this word is commonly used to describe marriage.

A Worldwide Death and Rebirth Ritual

A symbolic return to paradise lost is what this all equates to. Looking at much religious symbolism we can see that the fall of man corresponds with his split into two separate genders. This was the great punishment delved out to Man by god, duality itself. The only way out of this violent[103] state is to return to oneness. This is why the oneness meme is so popular within new age propaganda. We are slowly being introduced to the concept of *singularity*. Some, who are convinced they understand the meaning of this term, have yet to come to grips with its full scope. Its most esoteric interpretation is unfolding in time.

There is no shortage of esoteric symbolism when it comes to the actual creation of the New Man. The project to destroy and rebuild humanity is incredibly profound, but it is also a great horror. Make no mistake about it; this is a violent act. It is bloody. But then again, so is a birth! Here we see a diabolical justification, a justification used by many enlightened ones during their sales pitches for this monstrosity. Their enlightened perspective is this: The arrival of new life is violent, this is natural, and therefore we shouldn't worry when we see violence all around us. What we are witnessing is not bad, because it is bringing about new life. It is actually a birth.

Absolutely any idea can be justified through the intelligent use of words, symbols, and art. A powerful intellect knows exactly how to use such devices to justify its own actions, be they beneficial or malicious. The transhumanist project is being carried out by some very powerful minds. They know how to sell you their beautiful transformative ideas. They know how to break down the old order so that the new may rise.

The transhuman/ eugenic/ world political movement is a singularity. It represents the combined effort of powerful alchemists, known to the outside world as businessmen, politicians, military personnel, scientists, etc. They have taken it upon themselves to initiate and oversee a rapid evolution of this entire planet. This artificial evolution is itself an occult initiation ritual. Not just any ritual, but a mass ritual. In the past, similar rituals have marked important points in history, but this one in particular is of extreme significance. This is the One to herald their New Day. An entirely new age is being brought in. The new age itself is symbolized by the Phoenix, which rises out of its own ashes. The ashes of the old age are to be swept away as the fire of this great bird burns bright. An

---

[103] *Violence* being the supposed natural conflict between male and female.

enlightened age of reason will do away with the darkness[104] of the passing age. The old must die, so that the new may live. This is a massive death and rebirth ritual.

Yes, a ritual is nigh. The death and rebirth ritual itself is how individual candidates must be initiated into the Mysteries. It is the very first thing one does before becoming an Entered Apprentice (the first degree of Freemasonic initiation). Such rituals have always been the accepted method of becoming a sage, mystic, or adept. They are also a good way to make business connections. Such perks of Masonry are well known even amongst the profane. Ritual initiation introduces one to the inner world of the occult. Just as an individual is born again in this fashion, so can groups be reborn. When the neophyte is introduced to the altar, he is blindfolded. This symbolizes his ignorance. At the end of the ritual, the blindfold is torn off so that he may behold the Light. In this same way, the ignorant masses may be introduced to new ideas, and so be enlightened. At first new ideas are opposed, supposedly out of the ignorance of the mass, and not the fault of the ideas themselves. After the right culture-creation magic has been used, the mass mind learns to see things in a *new light*. This process is gradual but it is a powerful way to bring about changes in this world. It is the way to bring light to a dark world. This symbolism all points to our collective stupidity, and legitimizes the need for an intellectual elite. The enlightened ones rule by fiat, and they lay claim to power through sheer force of will. No doubt, their rituals work, but they are carried out beyond our conscious awareness.

Most macro-scale rituals happen over time. This is a simple logistics issue. Because of their breadth, mass rituals can't always occur instantaneously. Even large conflicts such as wars require some build up before a mass blood sacrifice can occur. War is a ritual? Yes, a very bloody and violent one. It involves much death, but from it *new life* emerges. It is on this point that we return to the symbolism of the ritual's birth. The birth that occurs here takes place in the mind of the initiate, and in the case of mass rituals we are talking about the mind of the collective. Through these rituals we are all being initiated into a change of ideas. We alter our behavior through occult rebirth. When it comes to warfare, entirely new societies are reborn out of the ashes of those that are conquered. People die to give life to an idea. This is all that government is; an idea. It only works when people

---

[104] The fire, or light of the new age is wisdom. The darkness is the ignorance of the passing age. This is a process of enlightenment (so we're told).

A Worldwide Death and Rebirth Ritual

agree with it. The extreme significance of mass acceptance is what inspires these rituals.

The death and rebirth ritual is how the individual initiate is born again, or born of fire. It is how he becomes a new man. His change in attitude is one level of occult meaning behind the term *new man*. The term Man itself is often used to describe the collective, not the individual. Man really means Mankind. This Man is to be returned to oneness with God. This is what the posthuman ideal is all about, and it also explains why transhumanists are always buzzing about the hive mind. The hive mind intelligence they talk about has everything to do with ultimate unification. It is concerned with the power of the collective mind. By harnessing a massive group of minds, the hive would have great power. Transhumanists explain this as a heavenly thing, a way to gain god-like intelligence. The dissolution of the individual is necessary to see this death and rebirth ritual through. It is the way that the dawn of the New Man arrives. Transhuman mystics want to see the literal death of Man (the human being as we know it) so that he may be reborn as the posthuman New Man. Their New Dawn would see their divine light made flesh. It is about transforming (transmuting) the dark mass into a brightly shining god on earth. A golden creation, that rises up from the base material of lead. This epitomizes the Latin phrase, E Pluribus Unum, "Out of Many, One." The subversive manipulation of the mass has been an age-old project, but now the stakes are raised. We are all about to be raised up from the coffin in a massive death and rebirth ritual to end them all. Don't worry though, because the death of the old man is a good thing. It will bring about a lovely new union of the world. This is the gist of the propaganda that we've all been hearing for quite some time now.

Mystic symbolism is everywhere to be heard coming from the mouths of those who know exactly what they are saying (even though you don't). For anyone not versed in the occult, such cryptic messages go unnoticed. When you clue in to their symbolism, the words of public figures take on a whole new meaning. Listen carefully to these words,

> I believe that with all the dislocation that we now experience, there also exists an extraordinary opportunity to form, for the first time in history, a truly global society carried up by the principle of interdependence, and if we act wisely, and with vision, I think we can look back to all this turmoil as the birth pangs of a more creative and better system.
>
> -Henry Kissinger, 1975

My, oh my, is that chock-full of buzzwords and symbolism. For one thing Kissinger is talking about his beloved New World Order, but again it must be understood what that order actually is. It is a collectivized "interdependent" world. La la la, it will be a happy place where we all hold hands and sing songs. Creating this paradise requires that we patiently trudge through the "turmoil" of our times (of which he has played a major role in creating). This turmoil has brought much death, both figurative and literal. In the high occult tradition, both sorts of death merge together. Through massive death, global society will be reborn. Kissinger's exact words for this were "birth pangs." I have found this exact same terminology (birth pangs) pop up in the speeches of many men who are calling for the very same future "global society," that Kissinger spoke of. A linguistic union is a powerful thing. All those who mention these "birth pangs" are referencing the upcoming birth of the New Man.

British Prime Minister Gordon Brown addressed the "international financial crisis" in a speech extolling the virtues of a world banking system. He brought out the "interdependence" buzzword, and much more for this one. He plays the same troubled times card that Kissinger did, because it is through the Hegelian process of *problem, reaction, solution* that global changes are made. Mr. Brown saw a light at the end of the tunnel,

> we could view the threats and challenges we face today as the difficult birth pangs of a New Global Order...   -Gordon Brown, 2009

The birth pangs are back. Tell mother it'll be all right.

Futurist, physicist, author, and major human enhancement spokesman Michio Kaku is yet another to utter these big words. In a "video podcast" he talked about science fiction, Internet, and the future of civilization. He said that we are going into a "type one civilization," a wonderful globalized world. The "transition" from our current civilization to a type one civilization would be difficult,

> and so this transition is perhaps the most important transition of all time. Some people don't want it. They fear this transition, because this transition is to a planetary civilization tolerant of many cultures. These are the terrists. In their gut they fear this, because they know they are witnessing the birth pangs of a beginning of a new planetary civilization, and the terrists want nothing to do with it,   -Michio Kaku, 2007(?)

In this lovely quote we get a dose of flowery peace and love propaganda alongside scare tactics to keep people from questioning

A Worldwide Death and Rebirth Ritual

it. Kaku says that the terrists (sounds just like terrorists) are the ones who fear this change. I believe that by terrists he means people who are attached to the Earth, or terra. This is similar to AI researcher Hugo de Garis's term "terrans," but the connection to terrorist cannot be denied. In fact he may be saying terrorists outright, in which case he is being quite brazen. It is hard to perfectly dissect his speech, but either way, this is pure propaganda. The "transition" buzzword is used too. *Trans*ition, as in *trans*form, *trans*cend, and *trans*human. To top it all off, he brings out the "birth pangs" line in reference to the mass death and rebirth ritual of the New Man.

One more example of the birth metaphor is seen in a February, 2003 TED talk given by biophysicist, author, and entrepreneur Gregory Stock. He mentions many of the transhuman advancements that are destined to come in the near future. He talks about drugs, reproduction, and preventative medicine. Near the end of the speech, he begins talking about how future generations will remember this point in time as a "glorious instant when we laid down the very foundations of their lives." This is quite symbolic, as it refers to the foundation on which the builders may lay their stones. Of course the Mystery Men are the builders, the Masons who create great temples. The sacred temple that Stock is talking about is the body itself,

> You know, it's a little bit like a birth where there's this bloody awful mess (that) happens, and then what comes out of it? New life, and actually, as pointed out earlier, we forget about all the struggle that was used in getting there,   -Gregory Stock, TED Talks, 2003

It is interesting to note that Stock slurs a bit and seems to catch himself while mentioning the "struggle that was used in getting there." He may be aware of the fact that violence and struggle is purposely fostered by the Great White Brotherhood to bring about their miraculous new births. Their radiant "new life" has a price. The death and rebirth ritual is a bloody awful mess, and a wonderful thing.

Fig. 9.1 The angels traversing the spiral staircase that is Jacob's Ladder.

A great idea is being conceived. To conceive is to bring to life. Babies are conceived, and so are ideas. The "big idea" which George H. W. Bush spoke about in his infamous September 11, 1991 New World Order speech is being born first in the mass mind. After it has risen to life within that intangible spirit realm, it may then be made manifest within this material reality. This is the process by which divinely inspired ideas are brought to the world. These ideas are brought by the angels, aka angles, the high initiates of the Masonic Order (among other orders), down to earth, the sphere of physical life. The philosophic use of the Masonic compass and square is all about this process. You use the square to measure the material, and you use the compass to bring your divine plan down from heaven (your own mind). From heaven to earth, this is Jacob's Ladder that the angels are constantly traversing. Up and down, down and up, this is the way the perfected society slowly takes shape. It is the way to eventually *trans*cend human limitation itself.

A Worldwide Death and Rebirth Ritual 195

Within mystic philosophy, the human being is a fallen creature that must be redeemed. He must be reborn into the Light of godhood. When we talk about the Fall of Man and his subsequent quest for redemption, we must understand who has taken on this quest. Who is it that has not only defined the great problem, but has also decided to create the solution? It is none other than the fallen angels themselves, those god-like men who come from heaven to earth for the sake of Mankind's redemption. They are spiritual beings incarnated within material bodies. You don't believe it? It doesn't matter, because they do. They identify themselves with divine principles. They have never cared about being human, because they identify themselves with something transcendent. They have fallen from higher realms into the lowly state of the material. The many weaknesses of their flesh are despised, because they hold back true creative potential and power. Their Great Work involves being redeemed to a state of perfect unification with god (their god). This Mystic belief system drives the entire alchemical project forward. The sublimation of the material world itself is their greatest labor. This is the Holy Grail, which will bring them eternal life. This is how they hope to perfect that which was left imperfect.

Everything that is being explained here could be expanded upon in great detail. Each powerful symbol relates to others, and forms a huge chain of philosophic unity. The mystery's Holy of Holies is the One. It is complete Unity. It can be portrayed as a lovely thing, but what does it mean to actually try and implement this ideal within our world? The compliance of huge amounts of people has to be manufactured in order to do this. You have to convince everyone that complete unity is good. More than this, you have to elevate the concept of unity to spiritual significance. If god is explained as one huge union, then people will surely want to be part of that. They will seek ultimate unity, but who will provide it? Is it god who is bringing unity to our world, or is it the fallen angels? Seriously, who is really in charge of this world? Massive blood sacrifices have been forced into existence through the directed efforts of a small group of financiers. These creatures must be understood for what they are. The veil that covers up true history must now be pulled away if we are to survive. Horribly violent events have been made to happen as a catalyst for change. This is what the common person must now come to realize. There are some people in this world that will do anything to get their way. Massive death and rebirth rituals have been planned and implemented by international financiers. This is how they gain greater degrees of control. Control over countries is gained by control over their money supplies. Where does our money come from? How is currency manifested in the first place? Who has the authority to create it?

Scientific advancement is funded with huge grants from governments, NGOs, and foundations. Huge corporations also fund research, many of which receive massive funds directly from the Pentagon. These are the companies that make the most significant advances in converging technologies. Quite simply, it is money that fuels scientific progress. The raw materials of the earth have been sublimated through commerce. This is how matter has been raised to higher levels of existence and value. This is the same act of rising that the death and rebirth ritual symbolizes. Metals have been mined and turned into more advanced tools to continue alchemical work. This is how society has grown in scale. International banking, politics, and science are all alchemy. They are technical instruments used to build up the New Jerusalem. The symbolic golden bricks used to build this holy city are actually human beings. Golden bricks are perfected ashlars. One attains such perfection through the death and rebirth ritual. This is how you become a perfected man. It is through continual death and rebirth. Within the life of a single initiate, many such rituals are performed. This repeated process is meant to mimic the seasons of nature, the continual rising and setting of the sun. The cycle of nature, the wheel of life, which is the completed zodiac, is present within a well-*rounded* individual. This all represents a natural process of growth and decay. Alchemy is about speeding up nature's slow process so that rapid evolution can occur.

Powerful men believe that they have the vision required to perfect the world around them. This is why govern*mental* and *think* tank white papers lay out scientific "visions" for the future. It also explains the symbolism of the eye in the capstone. This is the All Seeing Eye of Providence that shines down upon the world. The enlightened men of wisdom love all things to do with vision, because they believe that they embody the Light of god itself. Vision is the processing of light by the nervous system. The symbol of the light is very powerful indeed. It is physical light, radiation from the sun, which allows one to see. The Luciferian control over darkness (unthinking matter) through the use of light (intelligence) has brought our world to where it is today. Wise control of the world is symbolized in this way. After all, Lucifer is the Light Bearer. It is he who fell into this world with the light. The symbolism of the Mysteries makes it clear that one must seek the light in order to find his way through this world. This is the revealed light that occurs during ritual initiation. It is the light at the end of the tunnel that one sees after death. Those who don't have

A Worldwide Death and Rebirth Ritual

eyes to see never realize that the death spoken of here is merely metaphorical. It is the death of the ritual. The enlightened ones realize their power while they are actually living in this world rather than waiting for physical death to raise them. In this way they gain power over the masses which are forever waiting for their own demise.

Lucifer is an important figure in the death and rebirth ritual. He is known as the Morning Star, the light that heralds the new day. One interpretation of this is literal and scientific. It has to do with astronomy. The sun embodies the most ancient symbol of death and rebirth. Sunrise symbolizes birth, and sunset represents death. The actual Morning Star is the planet Venus; it is one of the seven great lights in heaven. Astronomically, these lights aren't stars, but they are referred to as such within the mystic tradition. They are the great travelers that move among the fixed signs of the zodiac. They are always changing their position in the sky (is it any surprise that politicians would revere them?) Venus/Lucifer is the first "star" seen in the evening, and the last to remain in the sky before the sun rises in the morning. This is the scientific aspect of Lucifer, which provides the material base for the philosophical identity of Lucifer. As the evening star, Lucifer brings about death, and as the daystar he initiates new life. The bringer of the dawn is bringing the Dawn of the New Day, otherwise known as the new age. This is the enlightened age of reason, and so on and so forth. This is the great symbolism of Lucifer. It is the lesser light that brings in the greater light of god, the sun. Here we see how Lucifer also symbolizes the enlightened man, the man whose Great Work is aimed at bringing in the light of God's Kingdom on earth. Luciferians see it as their place to oversee great and bloody rituals of death so that new life may emerge from them. They take on this responsibility because they view themselves as the most worthy of all men. They are material bodies imbued with the light of the Holy Spirit. They are gods on earth.

Light symbolizes intelligence. It is none other than the divine intelligence of god within themselves that Luciferians see as their great power. They believe that they are divinely inspired creatures. They are Lucifer. This too is part of their philosophical singularity.

```
           M
           A
           L
           E
                  FEMALE
    ━━━━━━━━┿━━━━━━━━━━━
            \
             \
              Orifice
            ┃
            ┃
```

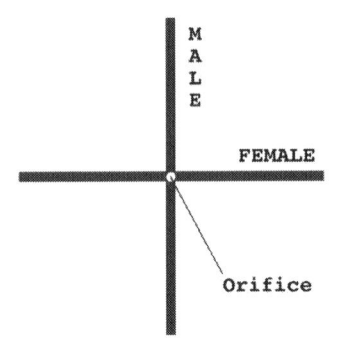

Fig. 9.2 The X-roads (intersexion) of male and female

Their spiritual control over the entire material world is self-replicating; it is an intelligent design.[105] They rule through the power of their perfected wisdom. Pure intellect is their god and the source of all their material riches. This is why intelligence, in all its forms, has been created by master alchemists. The creation of Intelligence Technology, Intelligence agencies, and Artificial Intelligence are all branches of the same tree. This is the literal Tree of Knowledge that the enlightened ones are not afraid to harvest. The forbidden fruit was bitten long ago and has since been processed for mass consumption.

Where we are at today is the product of wisdom, or Sophia, who is an interesting symbol herself. Sophia plays to androgynous themes of God the Father, and Mother Earth. Intelligence (symbolized by the male) imbued into this material world (which is the female); that is what Sophia is all about. It is the crossing of male and female principles that occurs in the quest to achieve material divinity. This quest is also explained in the ancient symbol of the cross. The actual point of intersection of the two lines (horizontal and vertical) of the cross is called the "orifice." This is where new life is *conceived*. This is an overt sexual reference, and it is of prime importance to the death and rebirth ritual. Everybody paint a cross on your forehead, the end of the world is coming! The massive death and rebirth ritual to bring in a new age involves the coming of a powerful light. Understand that this particular light is pure intellect. This is the divine power that will

---

[105] Intelligent design is typically attributed to God. The Masonic god is referred to as the Great Architect of the Universe, a designer.

breathe life into the New Man. That is assuming that such a future comes to pass.

The New Dawn project aims to magnify the power of intelligence, the very same intelligence that manifested on this planet long ago with early forms of life. Throughout time, intelligence has evolved, and a further evolution of intellect has become the focus of attention for alchemists everywhere. Master Alchemists now seek to bring life to a more immaculate being, a more powerful mind. The artificial intelligence, hive mind, interconnected oneness that everyone is talking about within the transhumanist community, has everything to do with this. These are all aspects of an esoteric ritual of mass initiation. Try telling that to a transhuman fundamentalist! They will have none of it, just as a fundamentalist Christian will become angry if you try to explain the broader truth of his religion, and let's not forget to mention all of those new agers who are always talking about "ascension" and working with "light." Ever wonder where these ideas came from? They came from Lucifer (the enlightened ones themselves). A singular ideology is being packaged for mass consumption under the guise of separate ideas. The new age is all about bringing light to a dark world. It is about initiating a hive mind collective intelligence through a massive death and rebirth ritual. The magnitude of this ritual is only properly understood by realizing that it involves the literal death of untold numbers of human beings. The old man, old order, old world, must be destroyed so that the New Man may be born.

We are being faithfully led to the end of life as we know it. Most people are actually looking forward to the end of the world, because the concept of a just rapture has been programmed into their minds. This idea takes whatever form that most fits each unique personality type. Sadly, the masses have been seduced into desiring Armageddon. They aren't concerned with understanding what the Apocalypse actually is; they would rather transmute their opinions into self-fulfilling prophecies. Prepackaged lies are quick and easy to swallow, whereas truth is hard to find, and even more difficult to face. It is a wonderful horror to behold and it is not for everyone. Because an honest yearning for truth exists in every one of us, and because most people are ready to accept false idols instead of diligently searching for real truth, the false light has become the most popular thing on the face of the earth. Yes, most people worship the false light, and coming to such an understanding is something to be taken seriously.

The reason why end times propaganda has taken many different forms is to facilitate the ritual initiation of the New Man. We are being made compliant in a project aimed at bringing about our ultimate demise. Understand that this ritual is to take place at the macrocosmic level. We aren't talking about one person. The death of humanity is all part of an enlightened plan to rebuild the entire world into a more immaculate creation. Man himself is to evolve with civilization. These two things, the individual human, and the greater society of which (s)he lives, are to become one thing: one god-like hive mind intelligence, one new gender-neutral collective of animated matter. A fiery rebirth would bring in a whole new macroscopic creature better in every way imaginable. This birth is really something to contemplate.

Light is prominent within the Mystery tradition with good reason. It represents the intelligent force of creation existing above physicality. Divine intelligence is the force that shapes matter "in the beginning." The different types of light recognized by mystics are varied, and interestingly enough, they all stem from an unseen and immortal force. The light of the Black Sun remains unseen. It is this invisible and intangible force that is actually worshipped as the highest spiritual principle. This form of light is the most immaculate. The most high priests of power worship a wonderful darkness. Those who see beyond the black and white, good and evil, male and female dialectic to be reborn as immortals gain true wisdom. Only through deep thought can such heights be reached. It is Lucifer who reaches toward the sun, only to fall back down to his lowly material estate, but that's OK, after all, this is how Jacob's Ladder works. If only this imperfect creation could be perfected, then the glory of god could truly be made manifest on earth.

As above so below: by bringing divine intelligence down to earth, the Alchemist could finally be redeemed to a state of spiritual oneness with the unseen creator. This was the task that Christ alone achieved. He too walked on water, which means that he existed within material reality while simultaneously transcending it. By realizing the unseen and divine aspect of his own light, he ascended to heaven. He realized that he was the Son of God, a material fragment of an infinite unseen Father. Because of this realization he was resurrected and he returned to the father. Both Lucifer and Christ are the models on which enlightened men base their own divinity.

By manifesting light in this world, the alchemists wish to gain true immortality. They literally want to realize godhood. For ages they have completed themselves through repeated death and rebirth initiation, which has brought them to higher and higher levels of intelligence. That is to say, by obtaining to higher degrees (their ranking system) within the secret society network, an initiate became privy to increasingly classified information. His level of light increased by gaining access to greater amounts of information. This is akin to the way that one gets involved in *black*-ops. A vast amount of information enabling societal control has been kept away from the eyes of the unworthy. This light is kept in the dark for good reason. It has always been the source of worldly power. Intelligence agencies are nothing more than secret societies. They filter out candidates based on their unique traits. The first and most important qualification for initiation is the ability to keep secrets. This is the fundamental darkness that sets apart all true-light workers from the mass of uninitiated crystal worshipping fools. True intelligence runs this world and it seeks to perfect itself. It wants to bring about its own death so that it may be purified, and reborn as a perfected new creature of brighter light. The interplay of literal and allegorical meaning within this symbolism is extremely subtle, and it is in this subtlety that the most profound realities of our world exist. The microcosm quickly explodes into the macrocosm. Anyone who lacks the eyes to see is blinded by the light. Accordingly, an immaculate and enlightened eye is used to watch us all. It is god's place to watch over us, to protect us. In a temporal world, a materialized god assumes this role.

Artificial Intelligence and all other converging technologies that are helping to bring about the birth of *god-like*[106] superintelligence are truly staggering to contemplate. The amount of change that could be brought about by Artificial General Intelligence is huge. Our entire world could be reborn by simply spawning AGI. Creating an AGI is one way that a man could give birth in and of his own devices (as in without a female). Such spiritual births are always the product of mind. This is how man transcends himself. He becomes a god by utilizing intelligent design. In this specific case we are talking about intelligence design, as in artificial intelligence technology. High priests of science understand this and they take their creative power seriously. Concepts, abstract and intangible spirits existing within the enlightened mind, lead to actual conception, or birth. Scientific study has the power to create new life.

We are already hearing transhumanists discuss the legal ramifications of defining "personhood" as it pertains to artificial life. Whether or not such life will exist appears to be a non-issue to many top AI researchers. In their opinion, it is merely a question of when

---

[106] This is the actual terminology used by transhumanists.

this will occur. Even before its birth (assuming it hasn't yet occurred), AGI is already being touted as superior to human intelligence. Our intellect is constantly degraded and lamented by those most eager to create AGI. If only a greater intelligence could be created, perhaps then we could begin to solve all of the problems in this world. This is an interesting rationalization, but what sort of change would superintelligence bring about? What kind of death and rebirth rituals would it take upon itself to oversee? As the greatest intelligence on earth, it would have the ability to lord over global civilization. All those existing under its rule, because of its superior intellect, may actually worship it. The inferior creator gods (AI scientists) already foresee this scenario. They are completely prepared to work with their new god in the hopes that it may grant them increased intelligence. Merging with AGI would be akin to walking into the light.

In an effort to attain godhood, death is considered a small price to pay. Many transhumanists speak about the pathetic state of humanity. We can see throughout their own literature that they truly do despise their own limitations. They hate being human. A personal death and rebirth, a transformation from human to posthuman would be a divine gift to them. So we see that death can be viewed as a gift. This gift embodies the great love that the enlightened ones have for us all. Because of their superior qualities, otherwise known as genetic endowments, their assumed responsibility is to bring about the death of all inferior types. In the grand scheme of things they actually see themselves as inferior to their cosmic god of light. This is the only thing they hold in higher regard to themselves. Both deep-seated envy and respect surround their relationship with pure intellect. Because intellect is the source of great power (thousands of years of lording over civilization has proved this point to them), they would do anything to attain greater levels of understanding. This power and the deep-rooted yearning to obtain it should never be denied by anyone outside the enlightened system.

Death is the greatest poison and also the greatest medicine in the alchemical tradition. Saturn, putrefaction, the color black, and the material state of existence itself are all symbols of death. The apocalyptic mass death and rebirth ritual is all about healing the sick and raising the dead. The dead are the unwashed masses who live in darkness without light. According to esoteric symbolism, they are dead their whole lives through because they do not have intelligence. One virulent form of sickness that the enlightened ones cannot stand is ignorance. All ignorance is to be done away with by bringing in the dawn of a true age of reason. If the inferior types can't be upgraded technologically, then they will be exterminated out of mercy. Such people have always been considered DOA anyway. Physical death would be a complement to their philosophical death.

Through all of this we can see a deeper reason behind the promotion of euthanasia and the grand eugenic plan. It is all part of a mass initiation into the Light of the Mysteries. The concept of genetic superiority is decidedly materialistic, and it should come as no surprise that it is an aspect of the enlightened one's esoteric religion. The pure blood of the Green Lion is that of the Natural Man (also known as the green man). It is the light of the sun that gives this blood its true power. Those divine natural men possess the inner spark, which sets them to the labor of the Great Work. This entire alchemical process is viewed as completely natural. It is through the intelligent utilization of human nature (intelligence) that the eventual attainment of godhood on earth is realized. It takes ages to complete this task. Continual death and rebirth, renewal of the material resources of the earth, is what is required. The lives of untold multitudes of human beings have contributed to this project, some knowingly, most unknowingly. The wise keepers of the Royal Secret have been those of the most pure spirit. The angels have reincarnated over and over again, have climbed up and down Jacob's Ladder, as they have forever labored toward an eternal throne. Actual immortality was the end sought. Interpreting the many allegories of the Mysteries can be done in a purely philosophical and intangible sense or a literal materialistic sense. It is impossible to say which of these two would be more correct. What is clear to see is an absolute hatred of duality. Unity is the Crown Jewel, the Philosopher's Stone to be had. The love of death is the great secret of the enlightened ones. Their rituals reflect the cycle, or circle of life. This circle is the zodiac. The One who exists above the 12 signs of the zodiac brings about the high esoteric number 13. Containing this wheel within one's self is the way to embody the eternal cycle; it is the answer to the mystery of perpetual motion.

The Power of Ideas

The world is directed according to ideas. Mind truly does exist over matter, but what would it mean to actually incarnate a supremely powerful mind onto this earth? The only way to answer this question is philosophically. We cannot do it scientifically and yet it is because of science that this light is being made manifest. Now do you understand the dramatic cohesion of Science, Philosophy, and

Religion? Are you beginning to understand that this unified system is broken up into pieces by design? The singularity that enables world control breaks itself up into a trinity. To initiates, the trinity is actually the number one. Pythagoras used the triangle/trinity to represent the number one. This is the nature of their triune god. It is divided up for mass consumption. This is done to fragment the minds of the people and keep them forever stuck in a flawed system of logic. The imperfection of this system is actually its perfection. The ignorance of the masses is the strength of the alchemist. Designed logic keeps most people stuck in a loop of thought, where all *answers* lead them back to where they started. The dialectic keeps them down and because of their supposed inferiority, they cannot see past this. They cannot go beyond good and evil, as the enlightened ones can. It is one thing to criticize someone who is stuck in this mind trap, but it is something entirely different to purposely design and perpetuate this flawed system of logic in the first place. There is a truly diabolical nature involved with maintaining this mind prison. We are all told what to believe in this world. Intimidation, peer pressure, and physical violence are all used to see that we conform to official party lines. Any party line, it doesn't matter which side we take on any given issue. There are many brands of popular thinking, all of which lead to the same place: death. None are really different from each other. Your downloaded opinions aren't as holy as you think they are. This is OK because you can make the decision to take out your own mental garbage at any given time. This is a worthwhile chore if ever there was one.

We must understand the power of ideas if we are to achieve the long forgotten prize of free thought. All of the secondary riches to be derived from that eternally elusive prize are barely known to the profane masses, but it is not because of their inherent faults that this is the case. People are not born to think in this way; they are educated as such. We all know truth naturally, but have it systematically beaten out of us our entire lives. This is the true death inflicted upon us all. We are all mentally and psychologically killed throughout our lives, never to be reborn of our own freewill. Any rebirth to be had is predetermined by our betters. They light our way. They initiate us into higher levels of consciousness. Monopoly in business is one thing, but monopoly over hearts and minds is far deeper.

So what exactly is the nature of ideas? Can they accurately be described as living things? It seems as though they can, and that is not my own opinion on the matter. Many scientists speak about concepts as though they are living things. In yet another TED talk from

A Worldwide Death and Rebirth Ritual 205

February, 2008 Dr. Susan Blackmore gave a presentation about *Memes and Temes*. Memes are ideas that are replicated: behavior, catch phrases, and pop-culture ideas spread by simple repetition. Blackmore likens this replication of ideas to the replication of genes, and in this way she gives a Darwinian explanation to the genesis of culture (culture creation). This is how she explains the science of memetics: culture evolves in the same way that biology does, but its deciding factor is memes not genes. Here, we find ourselves in a tangled mess of Social Darwinian rhetoric, as Blackmore proclaims that Darwinism is indeed "the best idea that anybody ever had." This is supposedly true, "because the idea was so simple, and yet it explains *all* design in the universe." That is an incredibly bold statement, a supreme act of faith. Properly explaining *all* life in the universe is not something that any science can claim to do completely, but that doesn't matter here. Blackmore has the stage, so she gets to say what she wants. The real thing to consider (and this goes for all other scientific presentations) is whether or not we decide to swallow what is being served up. Do we *choose* to replicate the memes presented?

Blackmore points out that the very word meme comes from the Greek root mimeme, which means: that which is imitated. True logic doesn't matter when it comes to cultural evolution; all that does is repetition. Examples of pointless memes are given as evidence that there is no rhyme or reason to human behavior. Monkey see, monkey do, as they say. In other words, we're all slaves to a mindless process (this is a subtle degradation of the entire audience) and a graphic explaining this is displayed on the large projector screen behind her. It explains,

> The Evolutionary Algorithm, or Universal Darwinism

which brings,

> Design out of Chaos Without the Aid of Mind

Clearly note that design is said to occur *without* mind; this is of absolute importance. We also see in this graphic a reference to the ancient Latin phrase *Ordo ab Chao*, or Order out of Chaos, which is revered in Freemasonry. Ordo ab Chao is one of the most important phrases within the Mystery tradition.[107] In Blackmore's graphic, the word order is replaced by design, but the two are interchangeable. A design is an ordered pattern created *deliberately* by mind. It takes

---

[107] Edward Bernays' *Propaganda* begins and ends with this phrase. Take a close look at the title of the first chapter, and the last sentence of the final chapter.

intent to create order, however the meme being reinforced here is that random change is indeed what causes evolution. This actually serves to reinforce the idea that there is no mind behind the madness of culture creation. We are supposed to believe that everything happens by chance. This could not be further from the truth. Anyone watching this TED talk who happens to have *eyes to see* will notice the cryptic reference to the ancient slogan Ordo ab Chao. Through this hidden message they will understand the real truth. That truth is that powerful intellects form and promote memes to the mass mind through popular culture. Ideas do not come out of nowhere; gods (which in this case are really humans) create them. Any fool who happens to be caught up in the purely scientific (godless) mindset will take what Blackmore says verbatim. They already deny the existence of god in any form, and thus perpetuate the foolish notion that all life evolves by chance. Their entire lives are thus made to look like the outcome of one large random accident. Our culture, which is explained as a product of memes (ideas), is somehow to be viewed as a completely random mutation. How exactly does that work?

The logical fallacies in this talk are one thing to marvel at, but their actual purpose is what needs to be confronted. These powerful memes have indeed taken hold in the minds of men. Cultural evolution has been carried out under a Social Darwinian model, but the true scope of Social Darwinism is not realized by close-minded fools. Flawed logic takes precedent, and because of this, the mass mind truly does take on a chaotic quality. You must understand the real power of ideas to follow what Blackmore gets into later. Interestingly enough, she says that,

> Language is a parasite we've adapted to.

She then explains "temes." Temes are "techno-memes," actual technologies that can be understood as ideas. Like memes, the temes replicate, and they influence the way that humans live and evolve. Blackmore describes temes as if they truly are living things.[108] In a sea of mindlessness, these temes will find a way to evolve on their own. This won't even require direct human thought or intent. In the sense that ideas have lives of their own, this is incredibly interesting. When temes evolve to a certain point, they will not only influence the direction of human evolution, but they will force humanity to adapt to their particular design. Blackmore insists that temes are:

> forcing our brains to be more like teme machines.

---

[108] Blackmore may have gotten this idea due to her background in parapsychology and the paranormal.

This is all put in the context of transhuman enhancement,

> We're going to have all kinds of implants, drugs that force us to stay awake all the time, we'll think we're choosing these things, but the temes are making us do it.

"The temes are making us do it." This really is something to think about. It builds on the premise of mindless evolution, while adding a whole new dynamic to it. Ideas and technologies (which combine into temes) actually influence our unconscious minds with purpose. Apparently, the machines have ambition, but lack intent. It is they who are rebuilding us in their image. From this point Blackmore pulls out the old "Is there anybody out there?" question. Of course she is referring to alien life, but in the context of living temes, this seems even more interesting. Perhaps the life "out there," we've heard so much about actually refers to the intangible realm of ideas. When these aliens manifest in our material reality they begin taking on a form of life that best suits them. These are the temes. Blackmore says,

> The temes are selfish replicators and they don't care about us, or our planet, or anything else. They're just information, why would they?
>
> -Susan Blackmore, TED Talks, 2008

It appears as if the temes are pure spirits, as they are beyond the knowledge of good and evil. Whatever may be the case, Blackmore certainly refers to them as living things. She explains them as mindless ideas that have an inherent need to evolve. In this task they are merely using humanity as a stepping stool to their own evolution.

If we are to take anything from this TED talk, it should be a wheelbarrow full of questions. Are temes really living things? What is life? What lies beyond, behind, or through life itself? Are human beings unconscious of the fact they are adapting to unseen living ideas? The concept of a thought-form is not new.[109] Ideas certainly have the power to bring specific form to living matter. This again is the dialectic of form and matter, male and female. The intangible and unseen Father is the spirit who forms the material Mother Earth. The thought-form made tangible is symbolized in Sophia, or wisdom. Sophia is female, because she is the material incarnation of divine principles. The perfected man, who contains Sophia within himself, is androgynous. He is a creator god. This is the perfection that the adept

---

[109] Theosophy was big on this. A movement that many influential abstract artists were deeply involved in. These artists sincerely believed they were in contact with something supernatural.

seeks; it is his rebirth. Taking this information and applying it to all of this Social Darwinian nonsense masquerading as science clears things up, if only somewhat. Ideas are believed to be alive. In fact it is the unseen Big Idea that is the most highly adored essence of the Mysteries.

All of the technologies now springing forth, seemingly out of nowhere, are believed to be divine incarnations. They are gifts from god that are aiding in the Alchemical quest of unification. Do they have a life of their own? Is there an unseen order within the realm of ideas, and if so how does it interact with our material world? It is not unreasonable to recognize purpose, and even a kind of life in that which lies beyond our sensory perception. Our world is certainly directed with purpose and yet we are made to believe that everything is random. If we believe this, then of course, we will automatically dismiss the notion that something exists beyond ourselves. It just couldn't be, based on the logic of science. All of this returns to our flawed system of logic, and how it is nothing more than a design. The geometric form of our logic is circular. It leads around and around, always returning to where it started. It leaves no room for actual critical thinking, because it is based on a set of assumptions. In this we are led to believe instead of to question.

We must be faithful if the power monopoly is to sustain itself. This is the religion aspect of the trinity that stands alongside science and philosophy. The "gods" of earth have effectively disarmed us by taking away our creative power, but what higher power are they in line with? Do they wish to unite with pure intellect, pure wisdom, pure spirit itself? Yes, they do. This is made clear in their entire alchemical/ scientific endeavor. They are Cosmists, otherwise known as Cabbalists, Sages, and Mystics. This is why they want to build AGI. More appropriately, this is why they want to breathe life into machines. They want to become true creator gods and give birth to higher forms of life. This is where the story gets truly interesting. Ideally, their creation will be superior to themselves. This is strangely and disturbingly appropriate. They may just be doing the will of an unseen intelligence that has existed for eternity. Incarnating this pure intellect on earth could surely bring about their own death, but it is a spiritual rebirth that they seek. That is the truly transcendent nature of this entire project, and it points to ultimate reality. The worldwide death and rebirth ritual will be the incarnation of their Divine Light. This will completely imbue spirit into matter. The divinity, with which they have always aligned themselves, may be brought down to earth.

A Worldwide Death and Rebirth Ritual

This is the Light rising on a new day. It is the ultimate ideal worth dying for. The material base of science is the altar upon which the grand candidate is being led blindfolded. As a macrocosmic blindfold, collective ignorance is a religion, a horrible faith. It is this faith, which will be swiftly torn away at the conclusion of the ritual. This will be the unveiling of the Light. It will be the collective rebirth of Man through the gift of perfect Gnosis. That is assuming we don't decide to rip the blindfold off ourselves, and run out of the temple before the ritual is complete.

-Chapter Ten-
Let's Get Real

The element of the unknown is without a doubt the most intriguing aspect surrounding the Singularity. Why wouldn't it be? The unknown is eternally appealing and elusive. There is no human way to understand what the Singularity would bring, but what else is new? All attempts to completely understand reality inevitably lead to disappointment. This is the power of the Mystery; the true and eternal Mystery that lies beyond all human interpretation of Sacred Geometry. What we face is none other than a collective struggle with the unknown. A profound consequence arises from simply labeling our situation a struggle; however, this label is sadly accurate. Operations here on earth have long been directed toward the discovery of a pre-determined solution to a grand problem. The dialectic has been tacked onto natural existence from the start, and as such we have been yearning for godhood ever since. The ancient allegory of paradise lost makes it seem as though divinity was once the state of man, and that this has been long forgotten in time. The high esoteric system that underlies all religion was designed to reflect this first cause.

With the power of freewill, self-righteous individuals have decided to crown themselves gods of this world. Science is nothing more than a means to their end, an end drawn up in the beginning. Dialectic materialism has separated us from our innate connection to eternity, and replaced it with one large struggle. This has been done to keep us forever running in circles as the gods complete their work of ages. Everything in our world has been explained in terms of conflict.

Since the dawn of civilization, conflict has been manufactured to keep populations weak and in need of protection. This is a profound realization to come to, although it is impossible to understand for anyone who remains trapped within the intelligently designed illusion. There is, without a doubt, a grand illusion placed before our eyes, and its power is truly something to behold. It is a mathematic design, a completely practical view of material existence. The establishment of rules, codes, laws, guidelines, statutes, mandates, and governance is all part of its calculated apparatus. Is it any wonder that law after law is now stacking up against us, against freedom itself? This is how tyranny is perfected. Safety and security build the psychological vehicle that we feel is necessary to drive on the road to freedom. Government is the DMV that issues our license. Permission to drive is granted by authority. Authority is a dangerous thing to bestow upon anyone or anything, as it is inevitably abused. We don't live in a free world. It's as simple as that. You can keep on justifying things in your mind in a vain attempt to prove this fact wrong, but why bother? It won't change anything. The box we live in is growing larger, but its fundamental structure remains the same. It grows by converging its massive system of rules into a single point. This is the Singularity, a super-dense black hole of codes, nodes, and bricks. The brick is the building block of the Stone Mason, the perfected ashlar. It has been squared on all sides. With perfected ashlars, a supposedly perfect society will be built. But this society will be perfect by what standard?

In the grand scheme of things, we can see that humanity is small. This realization can lead innumerable directions, and the contemplation of god is a perfectly natural reaction to such thought. Thought is an interesting thing for it is man's great gift, as well as the cause of his great sorrow. Open-ended questions have always perturbed those who need to know. An intelligent minority has always been aware of the true power of the mystery. That true power is this: when the infinite questions of life are left unanswered, true freedom reigns. All things are possible when one approaches existence in this way. When the mystery is solved, a directive force is established. Directors take the stage to lead us onward. Rules are drawn up that branch out from their solutions. A mighty oak tree is grown from a single acorn. Ideas are seeds planted in our minds.

Ironically, in a world full of scientific answers, confusion reigns. Under such conditions, we need to remember how to ask questions. Questions will eventually lead us to the source of our long forgotten freedom. It is surprising how much true knowledge can be attained by simply clearing the mind, as opposed to filling it with made-to-order thoughts. Quick and easy paths are hard to resist because they are the apparent ways to win our perceived cosmic race to the finish line. Have you ever wondered why we refer to ourselves as the human *race*? A race has a beginning and end point. Racetracks end up exactly where they begin. They are circles. So where is it that we are racing to? We might just be going nowhere fast. Without a doubt, the highest performance vehicles you'd ever like to see have been designed and tuned to win this race to Singularity. Team human

Let's Get Real 213

wants to take home that trophy, which is (r)evolution itself. We have to evolve as fast as possible, because slow rides are for losers. In a world of winners and losers, who is who? The winners are all those who choose to join the team and work toward winning the trophy. The losers are all those chumps who wait in line at the DMV just so they can drive to the track and pay money for a ticket to watch the show. What happens to that tiny minority who choose not to participate in this perverse delusion whatsoever? The people who don't care about winning or losing, but instead enjoy living, what happens to them?

All of the information contained in this book is useless if you cannot apply it in one way or another to your own existence. We truly must be able to live in the here and now and so, there has to be some practicality in these very words. Understanding that our world is filled with illusion is difficult, because we truly do need to be in touch with reality. So how do we transcend the illusion? How are we to make our way through this matrix we inhabit? How do we avoid the false choices laid before our feet? Why, oh why, does the world have to be this way? Well, the plain truth is that the world does not have to be this way. Things would be different if we wanted them to be. Simply understanding this fact is a great way to start truly living. It is amazing what can be accomplished when freewill and choice are engaged. This is the great power that we all possess and which has always been marginalized by those few who wish to own it for themselves. Don't ever be afraid to accept the responsibility of freewill. Never take it for granted or let anyone steal it away from you. It is your own divine right. Instead of buying into the grand lie, we can freely choose to be honest with ourselves. We can accept our shortcomings and gain real self-esteem. When we don't direct all of our energy toward pleasing someone else, we can finally find out who it is that we are. We can know ourselves. What a wonderfully elusive task this is. Recovering from all the abuse: physical, psychological, and spiritual, is difficult, but worth the effort.

Intense introspection is critical, and must be done before anything else. The world will continue to change, but the direction of that change will be influenced by the way we each choose to live our life. Continuing on the path of least resistance will only amplify the problems we have already. There is no such thing as a healthy quick fix. We must understand the cold hard facts, if anything is to be changed for the better. There is no avoiding this.

The very facts of our dire situation are steadfastly denied by those people who are convinced that they know everything. These folks will flat out deny certain information, because it does not fit within their intellectual toy box. Basically, this is what we have all been given to keep us in a state of mental childhood, a bunch of playthings, a plethora of attractively designed words and ideas that we can endlessly toss around the sandbox. Lovely words such as democracy, progress, interdependent, order, enhance, improve, and others have all been deployed as psychological weapons. They are pieced together scientifically, and in such a way that they spawn a completely believable virtual reality. Anything existing outside the pleasing confines of this virtual world will not compute. Our reasoning system will not allow its impossible existence. Ridiculous notions, actual ghosts, will take form and become real while cold hard facts go completely ignored because they don't fit into the design of our virtual reality. This is how the war of ideas is waged, but victory has not yet been declared. Until such time that virtual reality can maintain itself autonomously, and keep the masses completely content in their own ignorance then the struggle will continue.

The reality that most of us feel compelled to uphold is one of comfort. We want to remain comfortable. No one wants to rock the boat. Our flawed belief systems provide us with a false sense of security so we hold onto them dearly. We are told that we are living in the most scientifically advanced society that has ever existed. Because of this, we are free to enjoy our lives in the material splendor of the modern world. Medicine, travel, housing, and entertainment have been amplified, and we are much better off for it. Technological progress carries downsides of course, but ultimately our improved comfort and happiness make up for it. Science has brought about miracles, and it is poised to bring many more. If we keep progressing, then we will amplify all of the scientific advances already achieved and unlock more that remain hidden. The general idea of progress is forever given a positive connotation; this is the paradigm we are living under.

Actually, the modern scientific world has some serious problems, but they aren't the ones you'll read about in magazines. It is truly amazing to see the natural world completely exploited, as those who do so define their work as natural. These are the Dionysian Architects. Natural existence has to be completely demolished so that a replacement may be built and proclaimed natural. There is no way that anyone would ever put up with this situation, not if they understood it overtly. This is why we are made to believe that this unnatural system is completely natural. We have to perceive it as an evolution of human ingenuity. We must be made to love it. All historical records must be evidence to the fact that we live in a wonderfully civilized and progressive world. Meanwhile, true crimes against humanity and nature are perpetrated on a daily basis. For the very few of these atrocities that are actually recognized by the media, their explanations are nothing more than distractions. We are told that they occur out of some vaguely defined evil that persists in the world. The source of this evil is either pinned on a scapegoat, or left completely ambiguous. This is a simple diversion and it keeps the ball rolling. The true aggressor poses as the peacekeeper. It is the warlord who tells us that we need to bring democracy to the world, and he does this in the name of peace. We need to be peaceful as he continues his violence. We have to comply with his campaign of terror by accepting his bold-faced lies and cover stories. This reassures us because we get to believe that some sort of distant benefit will actually come from doing nothing. An interdependent world of peace is on the horizon. So feel good when that politician is giving his visionary speech about the future. He needs you to rally behind him to bring in the Light of the New Dawn.

Isn't it interesting that nearly everything we have heard about prehistoric humanity is presented in a negative context? When was the last time you heard anything about the virtues of this long lost era? Do you think, maybe, that our ancient ancestors led thoroughly healthy and spiritually fulfilling lives? Not if you've ever been to a museum, or read a library book. All you ever hear is that cavemen (men who live in the dark) died at twenty years old, and lived horribly up until that point. Elaborate stories have been constructed based on "fossil evidence" that lead us to believe that life prior to civilization was a complete horror, a constant struggle for survival. Everything is a struggle; we just can't catch a break in this great human race. These prehistoric stories are quite presumptuous if nothing else. How can we pretend to know what life was really like then? Is it possible that things may not have been as bad as we've been told they were? It is funny how events that predate history can be so clearly defined in the realm of ideas. You could say that the ancient world is now an officially sanctioned portion of history. The official story is pretty well accepted by most people. His story has been established in our minds, and not surprisingly it substantiates the claim that we are living at the height of human progress, and that our collective existence has forever been a struggle to survive.

We have to admit the true scale of our ignorance to understand how far we have been misled. The primrose path to utopia sure looks good, but that is because it must. Why would we bother with it if we were happy with what we had already? Strong willed people don't consent to slavery, and that is why so much psychological warfare is used against us. It is deployed so that we comply with this project. We must keep our feet moving as we chase that carrot on a string. Forever are we improving, and somehow everyone is depressed and neurotic. To mask this psychosis we all drug ourselves into delirium and faithfully act as though nothing is out of the ordinary. Convenient justification comes in the fact that people have always had problems, and we're lucky to have pharmaceuticals to help us out. Are you against science or something? How can you question modern medicine? Would you rather eat roots for twenty seven years only to die in a freezing cold cave? Yeah, I didn't think so. Just shut up, and like it already! This is the point of view of ninety percent of the population, give or take ten percent.

An eight-hour workday, a forty-hour week, some paid vacation, and a 401 K, does all this amount to true progress? Do we feel vindicated in the many luxuries that our modern lifestyle provides? Is the luxury of captivity enough for us to forget the freedom of the wild? Is there something that we would rather be doing if we had the choice? The answer to this last question does not have a universal answer. Many people are completely content in their mundane life, whereas others daydream about a better career, fame, fortune, etc. Not everyone is psychotic enough to actually enjoy being abused day in day out, and this is why we are given so many fantasies to chase. These flights of fancy keep our mind going in circles, as we constantly repeat "if only I had this, or did that," and on and on. If we want to get real, then we have to forget about these diversions. An unnatural world will spawn synthetic and useless alternatives to itself. The dissatisfied multitudes yearning for something better require direction. This is precisely where everything goes wrong for us. It is the directors who steer the direction of our natural impulse for freedom. They see to it that we don't ever do anything truly liberating. We are made to follow their templates for success, instead of empowering ourselves. We align to movements and ideas of mass appeal. This is where concepts, such as transhumanism, are seeded by respected and credible leaders. Men of letters show us the way to a brighter future.

We need to start questioning our professed leaders. Who are they, and more importantly, with whom are they working? Do they have any affiliations with special interest groups? Do they purposely keep their involvement in these organizations quiet? Who pays them to do what they do? The overall structure that finances popular ideas has to be unraveled from the top down. When one takes the time to actually do this, it becomes clear that many apparent leaders are merely front men for big money special interests. The fact that it takes much research to figure out exactly where all this money is flowing from should be a tip off as to how things really work. We aren't meant to see the big picture. Our puppet leaders are given the role of mythical heroes. Like Jesus, they show the flock the way to salvation. Popular opinion is a powerful force to be reckoned with. This fact is well known, and a premium price is set on its ownership. It is this price that is paid by those with the most resources. We need to start directing questions at the true directors, if we can ever get around to actually tracking them down.

Self-empowerment does not come from following the leader. It comes from deep introspection, and honest self-evaluation. We need to question everything, especially ourselves. If we don't constantly reevaluate our own condition, then we will be vulnerable to persuasion from others. We must remake ourselves; otherwise we will be led unwittingly to the death and rebirth ritual. Remember that the occult meaning of the death and rebirth ritual is self-improvement. This is its true power, and it fulfills a deeply rooted human need. We must now truly change things. One way or another, we must improve. Realize that the most powerful forces on this earth have a vested interest in your own personal improvement. They want to see that you are improved according to their standards. This will benefit them, not you. There certainly are greedy psychopaths out there, and don't think for a minute that they are all in prison or the cold hard streets. No, the wisest psychopaths know how to find power, and hold onto it. They know how to justify themselves; by appropriating the justice system that everyone else lives under and using it for their own purpose. They must remain above the law.

The time that we are living in is so important that few are able to come to grips with its deep meaning. Life as we know it hinges upon the decisions that we all choose for ourselves. The entire world is affected by the actions of those beings living within it. This is your great responsibility. It is a gift. You literally have the power to change the world. You also have the option of defaulting on everything that matters by ignoring this responsibility. If too many people choose to default, not only will the answers to important questions be determined by psychopathic control-freaks, but the questions themselves will be defined, and in many cases, deleted by them. Real questions will not be asked in the first place. It would be unlikely that anyone would even think to ask them. We face this situation today. Very important questions actually sound ridiculous to most people. We have been seduced into believing false answers. We have been made to enjoy completely artificial lives. Danger lies all around us and we can't even see it! Now is our chance to turn things around, by calling this entire charade into question. We have a tremendous opportunity before us, if we choose to reach for it.

One important question to ask ourselves is: do we really want freedom? Do we even know what it is? If we do decide to be honest, and really engage true freedom, then how does technology fit into the picture? Technology won't just go away, that is for sure, and we can't properly label it evil, either. We could very well choose to use technologies to empower ourselves. It could be possible for everyone to be self sufficient, free, and empowered through the use of technologies that assist in living their own chosen lives. The key word here is assist. Technology is a tool, and it can be used to help us out. A whole new dynamic is now arising, and it is something that most people take for granted. When the tool begins to integrate with the user, then the user slowly becomes dependent on the tool itself. Instead of fostering human independence, a state of dependence is created. A contemporary example of this can be witnessed in someone who finds himself out in public, completely lost without his *smart* phone. It appears as though the phone's intelligence has not rubbed off on its owner. He can't find his way back home without an internet map. This dependence on technology is a very real danger, and something that is actually being encouraged as we go further and further down the road. Is it possible that the producers of technology are actually happy about this dependence? Is the lovely interdependent world we have heard about from political big shots to be an evolution of this trend? Would completely integrating tools into our biology strengthen or weaken us?

Technology, we are told, is neutral. In a grand spiritual context, the truth of this matter may far exceed this base assumption, but for now let's go with it and assume that it is true. Technology is a completely emotionless and indifferent thing. It can be used for any purpose. Could it be that we are jealous of this? Do we envy the pure performance of technology and its ability to do things faster than us? Complete control over emotions is a very popular theme within transhumanism. In part this is due to a strange envy over the unfeeling genius of technology. Completely merging with emotionless technology is actually a high act of love for anyone of this mindset. The old compassionless love that the the Brotherhood of Saturn was so excited about is seen here in a different form. Human

Let's Get Real

love is typically marked by emotional extremes. This is a terrible thing from a purely practical point of view. Emotions and typical human love get in the way of pure performance. Transhumanists want greater power. The human mind, laden with its emotional baggage, just slows down development of this power. If we didn't get so caught up in emotion, then we could get down to business. Logic could finally be perfected in this way. This is a psychopathic dream come true: a world filled with people who are in perfect alignment with reason.

There is no way that the situation we find ourselves in will ever just disappear. Reality must be confronted, and it has to be done bravely. Some seriously difficult answers await all those who begin to question everything. If we choose to find freedom, then we must be prepared. Self-discovery might just reveal an ugly sight in the mirror. Upon such revelation, we may choose to face our own personal horror without flinching, or opt for the quick fix by taking a ride down to the plastic surgeon. Realize that what you see in the mirror may just be an illusion, so do not be afraid. Take charge of your personal illusion, and bend it back to reality. The quick and easy route only leads to destruction. You can remodel your face again and again, only to become a true-life monster. This is what cosmetic surgery addicts do to themselves, because they are unable to see the problems inherent in their own perceptions. See through the illusion and all will work out fine. Believe the illusion and you will become it.

We must get real in every aspect of our lives. There is absolutely no reason to fear this reality, no matter how twisted it has become. It is what it is, and no amount of wishful thinking will ever do away with it, so face up to it. Understand what is happening, and you will be ready. Don't ever sacrifice your spirit for someone else's benefit. Get real by constantly questioning everything. Keep your guard up for lies and distortions. Maintain *Eternal Vigilance*.

Internet Traps

A worthy criticism of this book might be that it is merely a collection of garbage, picked over and reused by a transient internet hack. This is worthy, not out of any fault inherent in these very words, but because so much of the truth that resides in cyberspace is mislabeled. To rely solely on the information of online truth dealers would be unwise. This is why I have gone directly to the root of the issues presented in this book by reading a wide variety of source material. The best information to be found comes directly from primary sources, and this is why I take the time to read what actual transhumanists, think tanks, and governments have to say.

To find the whole truth requires going beyond the closed circle of any intellectual community. This is the problem facing transhuman fundamentalists. They limit themselves to the transhumanist perspective, and so shut out other points of view. Dots have to be connected before one begins to see clearly. A serious obstacle to drawing the big picture is that many of these dots have been marked in invisible ink. We need to enhance our vision to be able to recognize them. It is only appropriate that transhuman enhancements are promising this very thing. In the most exoteric and literal sense, our limitations are to be overcome, as our collective ignorance is transmuted into brilliance. One has to wonder about the reality of such promises. Perhaps we should perfect ourselves by means already available before resorting to technological means. The enlightened bohemians who have created and are now promoting converging technologies believe that they have already done this. They believe they have achieved all that is humanly possible. Their intelligence has peaked and they are now ready to advance to the next level. This is why they are so eager to create artificial intelligence and performance-enhancing technologies. They are oh, so charitable in showing us all the light by promoting their project.

It is true that this book has been greatly inspired by the work of those within the alternative media, as well as other internet heroes, however we hold no single outlet or individual up as a definitive dealer of truth because there is no such thing. The best voices for truth on Internet are those that acknowledge the reality of seeking. It is something that the individual must do alone. Truth lies within the questions, not the answers. This is the nature of profound truth. Simple facts outlining our reality are trivial compared to this, but make no mistake about it, the multitudinous and unsettling facts of our collective situation are incredibly important now. With an abundance of information outlining our dire situation, the best voices of the truth movement have presented a challenge to us all. This challenge is to boldly accept these harsh realities, while remaining cautious of disinformation. That is to say, we must admit the validity of uncomfortable facts while remaining critical of all incoming information. Problems arise when any proclaimed truth is taken on faith. We must all investigate this evidence for ourselves in order to validate it. A lot of hard work is involved in collecting a solid batch of facts. The good news is there is so much information freely available through Internet that the process of waking up is now

Let's Get Real

quicker and easier than ever before. The bad news is that within this sea of information, lays some cleverly camouflaged misinformation/disinformation.

A huge mess occurs for most people during their internet wake-up call. An abundance of disinformation exists that has been packaged in very appealing fashions. Actual facts have been coupled with believable lies so that gullible seekers will latch onto them, as cynical skeptics debase the truth movement with the convenient label *conspiracy theory*. This is not good for anyone. Plain facts about the tyranny being built up all around us can be ignored through guilt by association. All of the true information and false disinformation gets lumped together by those who are either too lazy or too scared to search for the truth. As individuals we need to have the mental fortitude to endure all of this.

The many internet traps that have been set on the road to waking up are all too apparent. All of this beautiful disinformation is without a doubt intriguing. You may have seen it while pouring over posts on one of many Internet forums dedicated specifically to truth seeking. Much spiritually oriented content is served up there, most of which has a harsh aftertaste that the genuine seeker cannot force himself to stomach. Like aspartame, the chemical source of this bad taste must be revealed and understood for what it is.

Subjects such as aliens, 2012, Armageddon, and the mysteries have all been addressed and taken to ridiculous extremes all across the web. Some of the most outlandish conspiracy material is the most enticing. It plays to the mystery by providing an exciting answer to the unknown. Temptation leads many well-meaning people down false paths. By believing straight disinformation, one becomes disarmed. They can actually become part of the problem by reinforcing these false answers, and in this lays the real truth behind all disinformation campaigns. Disinfo is distributed for the sole purpose of continuing the chaos and confusion of the profane mind. It is actually produced to keep you in a state of ignorance. The keepers of the light do not want to give up their prize, and so they encourage the worship of false idols. This is why we cannot become know-it-alls, because all too often we don't really know much of anything. Instead we need to be question-it-alls. So let's get down to questioning a few of the suspicious answers to the great mysteries that have been strewn throughout cyberspace.

A very interesting batch of disinformation comes from channeled texts. Entire books (in many cases volumes) have taken form due to the supposed channeling of inter-dimensional spirits. It has been claimed that Human mediums have channeled the wisdom of these beings while in trance states. Channeled entities include Ra, the Cassiopeans, and other beings of light. The symbolism and cryptic messages in these texts reveal their true source, but the romantic message they convey appeals to passionate and well-meaning people. Unchecked emotion is one of three slayers that killed the Master Builder, Hiram Abiff. In this allegory lies some very important information. It is the untamed (and feminine) emotion existing within the heart of darkness that the Mystery Men truly hate. Anyone who exhibits this trait deserves to be fooled. They are weak. This is an aspect of esoteric philosophy that you absolutely must understand before buying into any disinfo. Realize that human emotion is not only despised, but also taken advantage of by the Light Bearers. Don't fall into their trap when they present you with uplifting tales about bringing light to a dark world. These texts are channeled all right, channeled down from the high-ranking spiritual authorities of this world, the initiates of various secret sects and military PSY-OPS divisions. These individuals aren't bothered by pesky emotion because they have risen above it.

A particularly telling and appropriate channeled text to bring up in regard to transhumanism is called *Bringers of the Dawn* (which I will refer to as BotD). This work's subtitle is *Teachings From the Pleadians*. It is alleged to be a work derived from the channeling done through Barbara Marciniak in the 1980s. The very title of this book is a cryptic declaration. Remember who the Bringers of the Dawn are? This book comes from individuals who are helping bring about the dawn of the new age. The messages contained in this text, and all others like it, are crafted specifically to get you in line with their divine plan. The method used to bend your freewill toward their ends is ingenious. *BotD* coaxes you into identifying yourself as a "being of light," a spiritual being working for the good of all mankind. You are told that you are actually a god incarnated in the physical world. In this way you actually become one of them, you become a willing "light worker." Once a huge dose of flattery is laid upon you, direction is given. All of that godly power you possess needs to be channeled in a useful direction. Light work, as described throughout new age propaganda, is the very labor that will bring in the New Dawn. This will bring the egalitarian world state of the enlightened ones, but you aren't supposed to understand the full implications of this. You are instead led to believe in an ideal that suits your own tastes so that you faithfully carry out the great work unconsciously. In this way, you actually fulfill a will external to your own, as you become an extension of the prevailing system. You become a tiny cell in a large body. This body is controlled by the brains existing within the head that sits at the very top of society.[110]

So what does *BotD* actually say? It speaks of a coming "age of light," which will bring about an evolution of humanity and indeed the planet as a whole. Higher levels of consciousness will be attained. This is to be achieved through the work of the "family of light." By simply reading this book you are identified as a member of this family, you are a "renegade." Doesn't it feel good to be special? It goes on to say that humanity is an experiment, and that the merging of cultures, creation of "new world orders," and foment of chaos and confusion are all needed to bring about this wonderful evolution. Very appropriately it says that "beings of light" are planning the future. How lovely, but what exactly is their plan? This is where things get very interesting, and the cryptic messages become clearer to see.

> What is occurring upon the planet now is the literal mutation of your physical body, for you are allowing it to be evolved to a point where it will be a computer that can house this information,
>
> -Barbara Marciniak, *Bringers of the Dawn*, 1992

If this isn't referring to the transmutation of man into cyborg then what is? For anyone who thought they could get away with interpreting this as a spiritual event, the reality is laid out,

> this is going to occur due to biogenetic engineering.

So be prepared. When advanced technology arrives promising you evolution, have no fear. Embrace it as a beautiful thing. We can now clearly see that the spiritual evolution found in new age propaganda is intimately connected with the transhuman agenda.

The philosophy of the enlightened ones is encoded within the pages of *BotD*. The point of this is to align the reader with the true family of light, as well as their divine plan. Marciniak writes about studies that are being done to see who has "recessive genes," in order to determine what individuals carry the "chord of light" within them. Remember eugenics? How about the Human Genome Project? Where, oh where, have our precious blood samples gotten off to, and why are DNA databases now a topic of interest in the media? Actual scientific experimentation is being referenced here. That is what is

---

[110] *BotD* when spoken sounds like *body*. This is probably just one of the infinite coincidences of our haphazard and pointless world.

truly amazing about this deceptive work. It actually tells some profound truths for anyone who is wise enough to see them. As a hereditary member of the family of light you may just be fortunate enough to be chosen by higher spiritual beings to bring in the dawn. That is the powerful esoteric message being conveyed. All those who are found to be worthy will be initiated. Part of this worthiness has to do with genetic inheritance, and that is why scientific tests are being performed to completely understand human genetics.

> You will have to demonstrate your integrity; you will have to pass an initiation or a testing to see whether you can be trusted with this kind of power,

It should be perfectly clear that what *BotD* is referring to here is the death and rebirth initiation rituals of the Mysteries. Appropriately, the macrocosmic death and rebirth ritual is outlined as well,

> It is a season when some things will die so that many new things can be born. It is all part of the divine plan.

Where have we heard this before? The heavenly plan is working itself out in time (the zodiac) so that new life may emerge in the proper season.

> It's an awesome task to carry light.

The great work, or labor, is always described as a privilege. No doubt, many people will be happy to take on the labor that has been dealt to them by their masters. The death of the old will not bother them in the slightest. These initiates will faithfully die so that they may be reborn. The technological means of attaining this new life are actually referenced here. As the text continues about "uniting consciousness," it also mentions technological implants. It actually describes a "five sided figure" that will be "implanted" in you. As a being of light you will choose to be "implanted." In other words, implants will be accepted as a freewill decision. Because of the promise of unlimited human potential, perfection, and even the ability to "fly" people will gladly accept implantation. These are exactly the kind of promises being doled out by transhumanism today.

Would you believe it if I told you that this book also speaks of the need to blend male and female aspects within the individual? Don't believe it, just read this book for yourself and understand what it is actually attempting to communicate to you. Read about how "life extension is coming back into fashion." Realize the true meaning of these statements and see the connection to what is happening today. Go beyond all new age propaganda. Be upset about the many deceptive tricks that are employed to get you in line with something

Let's Get Real

that you don't fully understand. See through all empty promises to recognize the true esoteric goal. Understand who is perpetrating this grand lie. The "beings of light," which have no problem fooling you for their proclaimed greater good, need to be held accountable.

> It creates a great deal of anger, unrest, havoc, and excitement in many human beings when they hear about undercover devices used to manipulate consciousness.

You're damn well right it does! Consciousness is the one thing we have left that may actually allow us to turn things around. Emotion isn't bad if it spurs you to truly inspired action. It is a wonderful human trait, which will be gone for good if you choose to implant yourself with the pentagram-shaped implant described in this book. All of this propaganda is aimed at disarming you psychologically. If you allow this to happen, then eventually complete annihilation of your survival instinct will be taken away via technological enhancements. At that point, psychological propaganda would no longer be necessary, and the human genome could be completely ironed out. Intelligently designed humans would be perfectly tuned to do the labor assigned to them.

Returning to the subject of internet forums, *Bringers of the Dawn* and other channeled texts have been praised by many people who see them as genuine. They have been taken as spiritual guidance material, and the act of channeling has been believed to be true in its most exoteric sense. People who consider themselves truth seekers have written long-winded odes and interpretations of these texts. Do not doubt the power of these channeled texts. Don't ever brush aside something that seems silly to you, because there may actually be a deeper story behind it. If you actually consider yourself a person who is dedicated to truth, then be sure to at least look at these things; just be sure to look twice. Constantly reevaluate your own conclusions. If you don't, then chances are you will fall for another form of propaganda crafted to fit your specific tastes. Although different in its superficial appearance, it will lead you to the same end as that which you mock. It will bring you to the Dawn of the New Day in which the human race evolves into something better and more immaculate. What a wonderful challenge it is for us to find the real truth.

The theme of a collective human evolution can be seen across Internet. All sorts of millennial raptures can be seen from the Mayan 2012 mythos to the technological Singularity of the transhumanists. Those who can recognize the invisible dots see the true singularity of these seemingly unrelated topics. For me, a chance trip to a popular bookstore brought about just such a revelation. While looking through the *New Age* section, this author picked up a book about 2012 that claimed to be a guide to succeeding in the coming task of spiritual evolution. The term *singularity* popped up from out of the many other words found on the back cover of the book and it was no accident. The word *singularity* is now a powerful meme that is being employed with direct purpose. There is a definite scientific and technological side to these apparently spiritual concepts. Science is the method by which all of these powerful ideals will be hardened into the material. The conjunction of the Alchemists is all about combination. A major line that one has to cross on the way to truth is the realization that the occult actually exists and influences our lives. The occult and the literal meet each other in the middle. They shake hands and do their work together. Progress is attained through literal science, and the mystery of the occult bamboozles people into believing a skewed version of reality. Perception is everything and that is why the dialectic has been unleashed upon us. Perpetual doublethink causes mental impotence.

What better way to psychologically disarm people, than by convincing them that the end of the world is coming? Armageddon, we've heard all about it. More and more people are beginning to believe that it is imminent, a foregone conclusion. Nothing could be further from the truth, for this is another skewed version of reality. Something horrible has been planned, but it won't bring about the end of the world. There is no such end, at least not as we understand it. This is another example of the psychological warfare that has been waged upon the unwashed masses. The desired result of this attack has been the formation of willing participants that will faithfully carry out the final stages of a plan. The "end of the world" merely represents an end to the way that the world is now. An entirely new system is being built upon the foundation of the old. You see the end is actually the beginning. Again, this is symbolized in the zodiac, the great wheel that brings regeneration with the passing of the seasons. Grand truth has been laid before you in signs and symbols. It is expected that you will not completely understand them. You are led to believe the half-truth of the Armageddon myth; that the end of the world is imminent. Your role in bringing about a beautiful new world is nothing more than death itself. We are being killed off on purpose. To make this process acceptable, psychological conditioning has been produced that makes us believe this is our fate. We are made to love it.

It all sounds so fantastic doesn't it? This just couldn't be. Some would call these very words disinformation. These criticisms are

Let's Get Real

bound to come and that is only a good thing. No truth stands alone. The words written here must be challenged, but the many insights must not be ignored. You need to heed these warnings and investigate their validity for yourself. You need to understand that far more is going on in this world than you have been made to realize. You need to get real and question everything.

The Dark Light of the New Age

The New Age movement has long advocated practical technological enhancement, and I believe that the transhuman merge of man and machine has always been its end goal. Any proclaimed spiritual benefits have been impostures. The cyborgization of the human race is in fact a major agenda and you will find that it pops up everywhere you look. From eugenics to the new age movement, the same idea is propagated; a collective global evolution brought about by man's enlightenment. A very interesting new age self-help book entitled *Saint Germain On Alchemy*, published in 1985 by the Summit University Press, offers some seriously important information for those who are wise enough to read between the lines.

Summit University was, and is, a destination in Malibu, California where retreats and classes are offered to people in search of new age enlightenment. The Summit Lighthouse describes itself as an outer organization of the Great White Brotherhood. It was founded by Mark L. Prophet in 1958 in Washington D.C. He was supposedly acting under the direction of the mysterious ascended master El Morya, who is the same allegorical character associated with the creation of the infamous Theosophical Society, a major organization that helped bring about the new age movement in the first place. Summit University claims to owe its existence to the "numberless numbers" of "saints robed in white."

Summit University appears to be like so many other new age destinations. It is a place where gullible people are parted from their money, but to leave its identity at this would not be appropriate. There is much more to this than most of us on the outside would care to realize. The book, *Saint Germain On Alchemy*, reveals much, and never specifically outlines any bogus supernatural rituals or practices that can give someone magical power. Instead, the reader is led onward, as secret societies are described as the best thing that ever happened to the world. The occult initiation rituals of secret societies are alluded to, but never named overtly. The job of the Summit Lighthouse seems to be the recruitment of adepts into appropriate divisions of the mystery school hierarchy. This book would certainly confuse many people in search of simple answers, which is very much its point. Of course many ridiculous exoteric interpretations can be derived from this book. The mythos surrounding Saint Germain himself, if taken literally, is totally ridiculous. It is implied that he was present all across the world over huge spans of time. He attended major historical events and was actually the man responsible for making them happen. As is the case with Francis Bacon, Saint Germain is a composite character. He is a symbol of secret societies and their many initiates who come together in common purpose. Their united effort is symbolized in characters that are depicted to the public as individual people. No doubt, many people who are given this book at Summit University take the exoteric fantasy as reality, and are thus easily led along a proclaimed spiritual path that will take them nowhere. For all those wise enough to understand the hidden truth, initiation into the mysteries is a very likely reward.

> What we are interested in is the subscribing of our students to a universal brotherhood and body already existing spiritually as the Great White Brotherhood. Being in the invisible, this order, comprising the alchemists of the spirit, requires a union with embodied humanity,
>
> -*Saint Germain on Alchemy*, 1985

The audience this book is written for is a fascinating subject itself. Anyone who studies cults knows that certain personality types are easily led astray by cult leaders. Gullible and desperate people who are searching for spiritual guidance can be made to believe anything. The new age movement has largely been about this sort of mind control. Traditional mind control techniques work all too well, but with advanced technology, they may be taken to all new levels of efficiency. This book mentions the dialectic process that leads our minds to conclusions. True alchemy is described appropriately. By focusing someone's attention in a certain direction, they will inevitably follow that direction. High alchemy has much to do with persuasion and it utilizes dialectic mind control. Adepts are simply those high level initiates who have gained control over matter. They have done this through a scientific method that can only increase in effectiveness as new age technologies are utilized.

Shortly after describing the New Man as "a golden man for the golden age," this text further describes some of the technological upgrades he will be equipped with,

Let's Get Real

> Let us consider for a moment the development of the mind switch... the mind switch is even more revolutionary for it will enable men to direct mechanical apparatus and electrical functions through brainwaves by the mastery of the energy currents flowing through the mind.[111]

This is obviously talking about brain-machine interfaces, which were not commonly discussed in 1985. BMIs are only now being released upon the consumer market, but they may have been around for much longer. If they weren't already available in 1985, then they were at least envisioned. Very specific uses for BMIs that echo those of the 2001 NBIC report are to be found here as well. Crime prevention will be carried out by a new and improved justice system,

> The wave patterns caused by criminal tendencies and crimes recorded in the etheric body will also be 'photographed'... evidence of guilt or innocence will thereby be afforded those administrators of justice who formerly relied on incomplete knowledge of events in the penalizing of delinquent individuals.[112]

This is all part of a divinely inspired government that will be established in the new age. It will hold its power over the populace in the name of science. Minds will literally be read, and no thought will go unrecorded. Funny how this is all being described as a lovely spiritual thing by the new age movement isn't it?

Of course Armageddon is described as well. The Age of Transitions, a time of catastrophic disaster, is always part of the new age program. It has to be, because it is in fact happening. As chaos arrives, we have to accept it as a natural occurrence so that we never search for the real reasons behind it,

> We are vulnerable to any number of potential disasters - economic collapse, nuclear war, crime, pollution, toxic wastes, and dangerous new technologies, not to mention the genetic engineering of the coming race.[113]

Here we see a description of the disaster-ridden end times coupled with a vision of genetically engineered posthumanity.

Another dramatic correlation to transhumanism can be seen in this book's description of human reproduction in the coming new age utopia,

> It is true that there are a number of studies being given attention at higher levels to alter the present system of giving birth, to cause the

---

[111] Prophet, *Saint Germain on Alchemy*, page 86

[112] Prophet, *Saint Germain on Alchemy*, page 87

[113] Prophet, *Saint Germain on Alchemy*, page 135

entire process to be painless and more immaculate, raising earth's evolutions into a new Christ era.[114]

This is talking about experiments with artificial wombs! It is describing their use as a way to become one step closer to god. Experiments in ectogenesis are described as projects that are being carried out at "higher levels." What do you think this means? It means that advanced sciences are developed in secret. Remember what the 2008 Harvard Science Review said? In that publication it was admitted that research on artificial wombs had been going on for a long time, and that it was purposely done in secret so that the public would not be aroused. Is it mere coincidence that this book about alchemy says the same thing?

One could write an entire book dedicated to interpreting *Saint Germain On Alchemy*'s true meaning, but let's leave things where they are for now. The truth of the matter goes far beyond any one book. If you begin to understand the true depth of what is going on in our world, then you will see the evidence all around you. Learn how to connect the invisible dots, and don't let anyone discourage you from attaining a true level of higher consciousness. It is interesting that the very thing that the new age movement offers to its cult members is freely available to anyone who has the gall to find it on their own.

---

[114] Prophet, *Saint Germain on Alchemy*, page 331

-Chapter Eleven-
Devil's Advocate

In so many ways, the Singularity represents something truly wonderful. Its complete benefits remain unknown and the mere contemplation of them leads an individual to desire what has not yet come to this world. Through this process, we witness the mystic quality that gives the Singularity its true power. It is alchemy. The idea being put forth is that Man could become a godlike being, recreated by his own intelligent design. He may at last be able to obtain immortality. This idea holds nearly irresistible appeal; it always has. Singularity is the Holy Grail itself. It represents complete Union, an esoteric ideal that has to do with actually becoming god. Microcosm and macrocosm converge at the crossroads of Singularity. In many ways, this represents the ultimate creative act, the highest form of sexual union. This is the more perfect union alluded to by the American founding fathers. It is an idea of massive appeal.

Society has grown by leaps and bounds over the past few centuries. The many revolutions: American, Industrial, etc., have extended the breadth of a worldwide civilization. To think that globalization has come about in our world only recently through some sort of accident or that it just happened as a secondary effect of progress is superficial and wrong. There is definite purpose behind *all* revolution, and history. We have to realize that history has actually been falsified to a great extent so that a powerful motivating force may be perpetuated. Myth is the occult identity of propaganda. Throughout time, myth has remained more important than truth because it spurs people to action. No one exists completely outside of the propaganda matrix. With this in mind, we have to remember to be kind to those individuals who invest themselves completely in this system. They believe that what they are doing is right. The players that we could easily point to as fitting within the structure of a grand conspiracy are likely duped, just like the rest of us. They believe they are right and that their version of history is true.

We all need to look deeper. The realm of ideas has to be further explored. We have to understand that our entire society, our world, has been built up according to a certain design. This design unfolds according to its own code, just like the seed that holds the genetic blueprint for a specific type of tree. Everyone within the framework of this design is operating on a need-to-know basis. We are all given our roles, and we play them well. Little do we realize the magnitude of the Shakespearean epic that we are all acting. This is because the truth lays hidden deep within the many Rosicrucian ciphers of the true Will-I-am Shakespeare himself. His design is not impossible to read; it simply remains encoded. One must simply decode the signs and symbols. Our real problem is that few people ever consider this as an option.

The mystique surrounding the Singularity is incredibly seductive. Its very essence is mystery, and as such it has the power to tug at our heartstrings. Its promise is just too great to resist, the possibility of uplift too wonderful. The awe-inspiring power of the unknown finds a powerful outlet in the form of the Singularity. This high ideal has become the rallying point upon which many different groups are now converging. As has been the case throughout human history, a powerful uniting idea has become the necessary catalyst for massive transmutation. This is how the Great Work of the Master Alchemist is achieved. It is born first in the mind and then conceived within the material sphere. Group psychology is used to bring about powerful transmutation. The actual force (Taurus) behind the change is the material world itself. This is the Apis Bull, the salt of the earth, otherwise known as the collective human population. A true beast of burden has been working hard for eternity, that is to say *we* have been slaving away for ages! It is our collective material body which carries out the alchemist's spiritual ideals to their literal conclusions. It is extremely telling that it takes an unthinking and unconscious mass to fulfill such spiritual ideas. This is the nature of the beast; its intelligence is inferior to that of Natural Man.

There is no doubt that the Mystery Men of the ages have been benevolent, at least within the framework of their own particular method. Each and every adept has sincerely believed their life's work was building up to a divine end. Arguing this fact is pointless. The real point to be taken from this lies deep within the elusive definitions of logic and reason. Their true defining quality is that of infinite possibility. Human potential derives all of its power from the mind of the individual. Our subjective reality is everything; it is the capstone of our existence. The enlightened eye, or I (I, representing the individuated human identity) upon the top of the pyramid is single. The implication of this is far-reaching. This Illuminist symbol

Fig. 11.1 Rubbings from Masonic tombstones.

communicates the overarching vision that stands above the structure of all human society. A singularity of purpose is the way to effectively build in this world. All those who have seen the light have aligned themselves to such higher purpose. They have joined with the collective Eye, the collective self. In the end, their own unique ideas about this project are of little or no consequence, because they become lost in time. What truly matters is that the work they did in their lifetime was actually that of the hidden hand[115] itself. The eye sees and the hand shapes. The historical myth envelopes the faithful initiate eternally. Their identity is forever grafted to the vine of the Mysteries.

A profound realization came to me one day while visiting a cemetery. Most of the graves surrounding me were lying quite humbly, flat within the ground and out of sight, but I couldn't help noticing some of the larger memorials standing high above the rest. These took the form of obelisks. Upon the face of these monuments I saw more than just names. Emblazoned alongside the identity of the deceased were various signs of the Craft (Fig. 11.1). Double headed eagles, pentagrams, and other symbols were cut into these stones. I was looking at the graves of Freemasons. The various degrees of attainment, the rank of the 32nd, 33rd, and so on were displayed next to each initiate's name. This was their distinguished rank in life, something they were incredibly proud of. Their degree was so cherished that it was literally taken with them to the grave. Beyond this, I could see that it was their own contribution to the Great Work and their affiliation with their fraternal order they wished to honor.

---

[115] In addition to his *Power Pyramid*, Mike Treder also wrote a piece titled *The Invisible Hand Needs Some Help*.

As wise men, they had always intended for this to outlive their mortal bodies. These men knew by contributing to such a large project, they could live on. A Masonic point of pride exists with the knowledge that one is working to build a better future for all Mankind. This is a core element of their belief system. In the hope that their contribution to the Great Work will be remembered and properly recognized when the light of the New Day arrives, they erect large tombstones. The true significance of these stones is that the death and rebirth ritual is inherent within them. The obelisks of these dying gods continue to hold creative power through their association with an ancient ideal. Osiris lives on.

We have to come to grips, every last one of us, with the fact that we don't know the whole story, we were never meant to. Even those Masons of the highest degree (33rd, exoterically) truly know nothing within the grand scheme of things. There are always Hidden Masters and Unknown Superiors at work. The Ancient Society of Free and Accepted Masons is not well understood by anyone who finds himself on the outside. This applies to the exterior facade of the fraternity itself, for it is merely the visible decoy of an invisible inner Mystery School. You see there is a difference between Free and Accepted Masons. They are not the same thing. Free-masons are *Born Free*, "as free as the wind blows, as free as the grass grows." These Freemen are the proven hereditary heirs of the Royal Secret. They are the true Dionysians. Accepted-masons are merely their glorified foot soldiers: men who are well educated, well connected, or otherwise initiated into the service of an exoteric fraternity, or working group of the Craft. In this way, Freemasonry is akin to a labor union. All of the best and brightest individuals who pop up throughout our society are properly trained and given high status. Not all of these gifted individuals are members of secret societies; many are simply intelligent people who find work within a certain field. The shaping of minds in our modern world is where the true power lies. The Mystery *School* has everything to do with education. Academia has become its public arm. The learning formula has increasingly been swayed toward specialization. It is through this process that we all lose some very important things. Not only does our practical knowledge of everyday survival degrade, but our ability to think is also altered. Our narrow focus makes it difficult to connect the dots. We have been divided, in oh, so many ways.

Not every leading transhumanist is an initiate, although it is highly likely that a good number of them have been accepted into the Craft. Please remember that most initiates (even those of high intellect) are not fully aware of what they are doing. Without a doubt, they believe their cause is just, and their affiliations beneficial. The type that the Unknown Superiors always look for in their ideal spokesperson is a liberal idealist. Individuals who believe that an Enlightened and intelligent government should be created so that the people of the world may be provided for, these are the ones chosen to spread the good word. Those who honestly believe that entitlements are the way to uplift people and who also have the intelligence and charisma necessary to sell a product; these individuals are sought out and used. The fact remains that throughout a lifetime of propaganda, most people end up aligning themselves to this progressive mindset. Taking our present system to its logical ends would bring about a benevolent and all-inclusive government that would oversee *every* aspect of our lives. It would provide every necessity of life: food, shelter, water, and more. A nanny state, a world where everything is provided for you, a society in which all that is asked in return is obedience and a small amount of your time dedicated to patriotic service. Who among us would take this offer if it were given?

The visible and apparent leaders of movements do their work with complete conviction. They honestly believe what they are doing is right. They know they have superiors, but they earnestly respect them, and don't see themselves as the victims of a pyramid scheme. In all likelihood, they bow and kneel before their superiors in some perverse power display ritual. If they do this, it is because they like it. Some people just love to serve and others get off on taking advantage of their servants. The incredibly sad thing is that this is the truth. Even presidents take their orders and they aren't coming from the people.

Words speak volumes. However, you must not simply read words, but understand their broad meanings to reach revelation. For instance, Mystic *Orders* of Fraternity are just that: orders. They follow a very specific and regimented sequence. We need to understand that beyond the superficial, there is the official. Master Alchemists are at work in this world, but they always remain hidden. This is the essence of their power and it permeates their entire creation. Everything they build acts as a protective moat. This is the essence of the fortress symbolism within Alchemy. The true powers in this world know how to hide behind their outer defenses. There is no shortage of well-known names at the ready to take a fall if need be. The human shield has long been used in this way. We think that the exterior, visible, temporal leaders that stand before us are the height of power. How shallow we are. Think about the structure of our very lives. We have all been so divided, compartmentalized, and specialized that we have become ignorant of everything beyond our own immediate circumference. We are all operating on a need-to-know basis. Consequently, we know nothing.

Regarding the occult Orders of initiation, it is well documented that high-ranking Freemason Albert Pike held a disdain for the lower degrees of the craft. In a letter to Robert Freke Gould, Pike wrote,

> I cannot conceive of anything that could have induced Ashmole, Mainwaring, and other men of their class to unite themselves with a lodge of working Masons, except this--that as Alchemists, Hermeticists, and Rosicrucians had no association of (their) own in England or Scotland, they joined the Masonic lodges in order to meet one another without being suspected.
>
> -Albert Pike's letter to Robert Freke Gould[116]

In this quote Pike is referring to the men who were actually behind the forming of the Royal Society. These were the true mystics living behind the legend of Francis Bacon. They were the true-to-life occult scientists. The creation of the modern accepted rites of Freemasonry (and the destruction of the history of its true origin) was nothing more than a needed first step toward their enlightened utopia. The real life New Atlantis would eventually have its citizens accepted through ritual initiation. All would be raised to higher levels of knowledge. It is exactly this project that we see at work in the University setting of today. The graduation ceremony is actually a ritual initiation for the profane. The college itself is the perfect way to direct candidates to established degrees. This creates an efficient division of labor. Universities are the exoteric/external schools that exist because of the impetus of an inner/hidden school. We are never meant to recognize the full scope of what we are doing. This reveals the deceptive nature of this enlightenment project. If we are being enlightened, then why don't we understand what we are a part of? Why aren't we told about the inner school of the mysteries?

> This school of wisdom has been forever most secretly hidden from the world, because it is invisible and submissive solely to Divine Government...[117]

This was written during the Enlightenment period by Karl von Eckartshausen in his *The Cloud Upon the Sanctuary*. He continues,

---

[116] Hall, *Masonic Orders of Fraternity*, page 263

[117] von Eckartshausen, *The Cloud Upon the Sanctuary*

...it is the society whose members form a theocratic republic, which one day will be the Regent Mother of the whole World.

Regent Mother, or nanny, is the one who will not only provide for all of us, but actually give birth to us. The family unit itself cannot stand up to the pressures of the artificial womb, genetic engineering, and life extension. These are all aspects of one large plan that is to come to fruition in the new age.

R.C.W. Ettinger, an early figure within the cryonics movement, was sincerely concerned about the future of the family unit. He said that in some form, family must persist, but he also understood the dramatic implications of engineering "supermen." He brings up fraternity when speaking of the transhuman world of the future. All those that will emerge from their cryostasis, who are thawed out and brought back to life, will experience a new kind of family,

> The kinship between ourselves and our historical parents -- if they also survive to become supermen -- will be fraternal instead of parental.
>
> -R.C.W. Ettinger, *Man Into Superman*, 1974

The family was never meant to make it through this Alchemical process. In its place, fraternity will provide a common bond between posthumans. Regent Mother was a very appropriate use of words, and men like von Eckartshausen choose their words wisely. The state will usurp Nature's role as the All-Mother. Behind the physical body of state, the mind will exist. High intelligence will guide this system. The gods of Divine Government will continue to be the brains of this operation, and their creative powers will grow to the height of the true god that they killed long ago. At least, this is their ideal. What will happen remains to be seen.

God is dead. This is a highly charged phrase. It is actually the mantra of the inner esoteric orders themselves. Masons are known as Widow's Sons. A widow is a woman whose husband has died. Thus, Widow's Sons have a dead father. Their mother is Isis, the mother of the Mysteries. Isis is the living material world itself, otherwise known as Nature. So we see that Isis is Mother Nature, mother to the Natural Man, the initiate of the mysteries. God, the Father is dead to them. In one sense this actually refers to them! The dying god, Osiris, is Man. As a mortal he dies, but this mortality is conquered through sexual regeneration. In another sense, *God is dead* also refers to the father of

238   **Revolve**

Fig. 11.2 Friedrich Nietszche as a young man showing off his Lion's Paw, a sign of initiation among brothers. Nihilism is but a reflection of the greater Ancient Mysteries.

Osiris, who is Saturn. Saturn represents the cold black void. He is the dead material world; that of mineral, of stone itself. It is upon this stone that life mysteriously incarnated. Life is the light that defies this darkness. It is the great rebel. There is no god existing above the Enlightened Man himself. He is the embodiment of light, the perfected stone. This is the essence of the alchemical process; it begins and ends with death. Death is what gives the initiate new life; it is how one is reborn. Life is a thing to be reshaped by its own wisdom for it is the Light. To join light and dark is the high spiritual quest of the initiate. It is the method by which one defies death. This is the esoteric belief system, the divinity that they have bestowed upon themselves. Since they are the brightest lights, they have the duty to improve all dark aspects of Nature.

There are definite reasons for everything that exists within this world. Align yourself to this truth and you will be far better off. Life has a purpose, and so do we. Get in touch with your own unique purpose and you will have earned your own life back from the master thieves. Hermes is a master thief and the Hermeticists are just that. Realize too, that Hermes is the god of commerce. So we see that the Lord giveth and he taketh away. He is a criminal mastermind and a control-freak. You will witness the dialectic fall in on itself when you view all of its angles simultaneously to reveal the big picture. How does it make you feel to realize that a small portion of the population eagerly anticipate an end to the family unit? Where do you think corruption stems from, and how could it ever go away?

We don't need to be told what to do, because we are strong. We don't need anyone else's perfection, because we already have our own. We need to understand these things now or we will continue to be the victims of incremental transmutation. We need to start asking ourselves the hard questions about our current state of affairs, never mind the future. What have we become already?

Twisted Virtue

The only way this system gets away with its tyranny is through our willing compliance. It is exactly this compliance that is manufactured through propaganda. Our unconscious minds and instinctual drives are being taken advantage of. The only way out of this mess is to attain conscious self-awareness. Given the difficulty of this task, we can see that something is horribly wrong in our world. We have to struggle to be ourselves. What kind of freedom is this?

No one can doubt the intelligence of this system. Its true power comes from its knowledge of how to exploit human virtue. Every wonderful attribute that defines our species has been scientifically analyzed with precision. This has been done so that our very nature may be controlled and pointed in a certain direction. Our psychology is deeply understood by the Master Alchemists. However, it is not enough for them to simply understand our motivations, instincts, and emotions, for they must also manipulate them. Our instinctive drives are set on track by these control freaks. They know what we want and make sure to tell us how to get it. We never see that we are being led to false promises. Little do we realize that by following their lead, we end up right where they want us. We are unconsciously fulfilling a directive that we don't even know exists. It is literal mind control that we are under.

Our natural instincts are being redirected. Because it is our virtue that is being exploited, we believe that everything is all right. We are honestly pursuing goals that we feel are honorable, and so a false sense of security arises. Only through deep introspection can we come to see the true source of our motivations. What we believe to be freewill decisions are actually instilled programs. The problem is not that we lack freewill, but that it is being usurped. Our choices are being programmed systematically, the same way that computers are programmed to carry out specific functions.

One very important virtue is that of self-sacrifice, or duty to the community. We naturally put the health of the tribe before that of ourselves. This is actually a survival instinct that keeps the human family alive. One can see the importance of the human bond that transcends the individual. It goes on after we die and while we live, it extends far beyond ourselves. Through community, our lives are enriched. We get to be a part of something greater than our self, and this is good. People choose to work together because it is so much better than being alone. This is natural behavior that helps us survive.

The natural inclination toward self-sacrifice has forever been exploited. This has been done to redirect the power of the collective toward ends that benefit a selfish few. This is the essence of the system that surrounds us now. It fulfills its own needs while simultaneously claiming to serve ours. Sure, we have running water, heat, and many other conveniences, but what is the price? How much effort are we putting into this system, and what is the return? There is a reason why a tiny minority of the population has become extremely wealthy, while leaving the vast majority to work their lives away only to attain a relatively meager level of subsistence. We are all slaving away, and the worst part is that we aren't given any alternatives. There is nowhere to go, because the system has been integrated into every aspect of our lives. The hopelessness of this monstrosity is parlayed as we are told that it actually benefits us. By dedicating our lives to this beast, we are doing a favor to future generations, or so we're told.

Our instinctual service to the species has been redirected toward blind service to the government. We have come to trust government as our benefactor and protector. We gladly work with it, and for it, out of our natural need to protect the collective. Through patriotic propaganda, we choose to go along with the program no matter what. We believe that our military's main objective is to protect us, even while it is being used to destroy innocent people across the globe. In a sick way, we actually feel secure while this goes on, and we naively expect that the high-tech weaponry of the Pentagon will never be turned against us. It already has been! We have become so dependent on the security provided by government that we never take the time to see it is an illusion. We don't understand the only real protection in this world comes from within ourselves. By being powerful individuals, we all protect each other. Our wellbeing doesn't depend on any sort of governmental structure; it merely depends on our own goodwill. Sadly, we have been inundated with prefabricated organizations that fulfill our need to help one another. All kinds of programs, foundations, and charities exist for us to join. We don't know much about these organizations, and we aren't supposed to. We are meant to fulfill our primitive drives without questioning their outlets. What are the true goals of the many self-proclaimed benevolent institutions that are everywhere around us? Who owns and operates them?

We need to ensure that people are provided for! This is the rallying cry. In our rush to fulfill the presumed needs of healthcare, education, and all of our other political emergencies, we have forgotten how to feed ourselves. We literally have been separated from our own food. All of the elements of life that used to be procured by the individual are now conveniently provided by services. This puts us in a state of dependence, and that's exactly the point. We aren't being empowered by this system. We are being disabled so that government may take over the role of the provider, the father figure. This is extremely dangerous. Our instinct toward self-sacrifice is being redirected toward mere worship of this provider. In a healthy world, our instinct would be fulfilled when we provide to others according to what we have to offer. An empowered individual empowers his community by sharing. Seeing that we no longer have anything of our own to offer, we must join up with those who have resources. We become human resources in a chain of codependence.

The need to improve, to be a strong individual, is also being flipped on its head. The virtuous task of self-improvement is actually the defining quality of transhumanism. The idea is to improve, to be a stronger, wiser, and all-around better person, to be "better than well." No doubt the transhumanist project has set out to make things better, but how so? Are the many technological upgrades they speak of actually improvements, and if they are, then what do they improve? How is human performance being enhanced? Again, we run into the problem of subjectivity. *Better* has to be defined. Standardization rears its ugly head again. By attaining to a higher standard, we could honestly say that we have been improved. But who is setting the standard of improvement? What is it that we are doing better? What sort of intelligence is being enhanced and for what purpose?

There are infinite ways that we could go about upgrading ourselves. To say that our own personal enhancement will take a path that we have sole control over is not true. If we decide to put a microchip in our head, then we are putting complete faith in the designer of that chip. We will inevitably find ourselves at the mercy of his design. The development of these technologies has required decades of research and development, not to mention trillion dollar annual budgets. It is the culmination of military, academic, and corporate partnership. It is an integrated project of government. To think that the technologies provided by these sources won't be designed to benefit themselves would be foolish. We need to understand what the true interests of these massive entities actually are. The role of the individual has to be put into the proper context. As it stands now, individuals represent a microcosmic portion of the institution itself. Our role is that of a tiny cell in a larger organism. Our entire lives are directed toward sustaining the good health of the corporation/ institution that employs us. This is why corporations have been legally defined as persons, because that is what they are! We have become nothing more than a speck within such bodies. This metaphor is actually used in the NBIC report during some speculation on the future,

> monolith corporations with cells of individuals who can do tasks, and as those tasks move from corporation to corporation, the cells would move as well.[118]

Our desire for self-improvement is a good thing, but it has been skewed. Our bodies are worth money and our minds even more. It is the mind that tells the body what to do. When our minds have been aligned to the service of someone else's interest, then we become profitable to them. Of course, we expect something in return for this effort, and our employer provides us with a certain wage. The deep question to ask here is what does this compensation amount to? The distribution of wealth in our country (and the world) has continually favored a smaller and smaller super-rich elite. Our work is creating huge amounts of wealth, but where has it all gone? We are working for corporations, but they are not working for us. Above these corporations, exist the finance oligarchs themselves, the Gold Men who funnel huge amounts of "bailout" money into their own private pockets. How can we sit by and quietly watch this? How can we continue working away as criminal activity takes place at such appalling levels?

---

[118] *NBIC*, page 271

Our behavior is already automated to such a degree that nothing seems able to steer it off course. It is appropriate that in a world of already robotic humans, the high posthuman ideal is being popularized. The transhumanist vision is nothing more than an upgrade to the current system we live in today. We feel the need to be faster, to work harder, to be better, but we don't ever take time out of our day to realize what it is that we are doing in the first place. In many ways, transhuman enhancement is a quick fix to a massive problem. By simply adding better components to our bodies, we will supposedly be better ourselves. A troublesome detail lies in the fact that this process of self-transformation doesn't really require much effort from us personally. All we need to do is find a way to obtain these enhancements. Either by buying them or having them secured through some form of government entitlement, we will end up winning. One can see the fault in this formula. Very little is required of the individual beyond mere faith. Improvement will come upon application and so the goal becomes to merely apply these technologies, and let the magic happen. Actual self-realization is not required; rather it is supposed that such a thing will be the byproduct of enhancement itself. Transhumanists want to be better, but they faithfully believe that this will occur due to an idol that exists beyond the self.

The collective yearning for peace was explained earlier, but the full magnitude of the world peace project may have been missed. What is extremely disturbing is its biological aspect. The actual physical removal of aggression from the human body is being contemplated on scientific terms. If we are to be peaceful, then there will no longer be a need for our animalistic aggression. Aggression itself is seen as a hindrance to sustaining a peaceful world. In his *Cosmist Manifesto* AI developer, Ben Goertzel, explains that the human mind itself may be too vicious for its own good,

> If so, then the only solution to making a happy society -- to really obsoleting the social dilemmas -- is to modify the human mind/brain... solving the problems of society at the source.

According to this worldview, we are the problem, or more specifically, our brains in their present form are the problem.

Our human instinct toward aggression is not solely a bad thing. It is all about the way in which that aggression is directed. In this sense, our virtue has been falsely labeled and scheduled for elimination. Survival depends upon one's ability to fight back when necessary. It appears as though this very instinct has come between the crosshairs, and is it any surprise? This fits in neatly with so many other elements that are now converging to bring about peace. In the name of safety and security, we are being led to the belief that we are a danger to ourselves. All things natural are being demonized, while the artificial becomes the celebrated solution.

If you want to see the drawstring that is bringing everything together, the real catalyst for convergence, it is the worldwide promotion of so-called safety and security. The *security* meme has been taken to epic proportions in our Post 9/11 World. Across the board, we are being led to a more secure future. This is why we are seeing the rise of governmental departments such as the Department of Homeland Security, but it doesn't end there. Everything is now tying in to national security. What does this mean for you and me? In the most profound sense, we are being made to fear ourselves. The object of terror, of our hatred, is slowly but surely shifting away from an elusive Middle Eastern devil man, and toward the man next door. The *homegrown terrorism* meme is now rising. The danger of this trend cannot by underestimated. A militarized police state is being established domestically, and its main target is you.

We are being convinced, little by little, that we are a danger to ourselves. Propaganda crafted to dehumanize us is everywhere we turn and it really deserves more space to detail, but its essence is that of self-loathing. The general idea is that humans have caused a horrible mess for themselves; the very mess that we are in today. Therefore humans become the target of elimination. The world would be healthier if there were fewer of us. Many people now believe this. A disturbing new faith is arriving and as always, it centers on fear and guilt. You would think that by now we would see this one coming. What kind of improvements do you think will be created in an environment of self-loathing and paranoia? This is a question I would rather not find an answer to. Instead, let us call out the real threats to our security.

Quite frankly, if you can't see through the illusion of the War on Terror, then you are delusional, or willfully ignorant. More than enough evidence exists to prove that the official story is filled with holes. Not only 9/11, but most other major acts of terrorism have been made to happen on purpose, thanks to the aid of intelligence agencies, so that the public would more strongly believe in the phony safety and security offered by government. The already massive legal system is transforming into a tyrannical apparatus of complete control, and it is not by accident. Through the promise of safety and security, we are being made to accept the unacceptable.

The genuine way to security is through an empowered and free population. Strong people know how to take care of each other. We have all been psychologically disarmed, debilitated, and reduced to an unnatural state of weakness. We are most certainly vulnerable to attack, but who is on the offensive? Danger does exist all around us, and we must realize that it is coming from institutions that claim to exist for our benefit. Despair has been cultivated within us so that we will beg for false security. Unless we get real and understand that true safety begins and ends with us, then tyranny will reign. Many weak-minded fools will sign up for government jobs to "protect" their fellow citizens. These poor souls will actually patrol the streets and punish their fellow slaves for violations of policy while their masters gladly suck the life out of all of us. We are being exploited to the utmost and now is the time to realize it. Understand that technology is being developed by the world's most powerful psychopaths to perfect their covert system of control, not to empower us.

The threat of Al Qaeda is actually used by transhumanists in their campaign for technological uplift. This is so predictable that it barely requires mentioning. What is more interesting is how the safety and security meme is applied to emerging technologies. Security is among the many benefits of these technologies. Citizen spying programs and advanced AI pattern mining systems are sold as a great way to detect terrorists. As privacy is completely eliminated, we will all gain the benefit of complete knowledge sharing. Ben Goertzel claims that this kind of sousveillance is actually a good thing. It will allow us to "watch the watchers." Given the levels to which government has become increasingly secretive, this doesn't seem like an accurate forecast. What seems more likely is that massive surveillance systems are being built to monitor the masses. All personal technologies, those that are carried, worn, or implanted, will feed data to these systems. This is all part of their functionality.

Beyond the Big Brother panopticon, there exist further definitions of safety and security that apply to transhumanism. Even before their upcoming public debut, artificial wombs are being sold as a safer way to birth children. This ties in to the health care system and its parent, the insurance industry. It is through the insurance companies that new standards of safety will be adopted and enforced. All medical practices deemed unsafe will exist outside of the evolving healthcare system. Natural birth could get phased out merely because of its cost. Money is now affecting human evolution dramatically. We need to get in touch with this fact and understand the ramifications.

Another feature of the evolving healthcare system is biosurveillance. By placing sensors in our bodies to monitor our vital signs, we could be treated more effectively. We would be safe, because response time to emergencies could be immediate. By unraveling everyone's unique genome, health practitioners would gain the ability to know which drugs work best for each person; this is the concept of pharmacogenomics. All allergies would be known immediately, and the best remedy would be applied according to genetics. Privacy in the world of medicine is a hindrance to any such changes. This is mentioned in every single policy paper written on the subject. Laws have to change and so do ideas. Our idea of privacy is shifting so that we will faithfully hand over increased control of our lives to our perceived provider. The big problem here is faith. We must have faith that this system exists for our benefit.

"Smart homes" and other autonomous systems may be the technologies by which our medical services are applied. Such homes could monitor everything that we do. They could also watch us to make sure that we are being safe. Implanted or otherwise internal biosurveillance systems that chart our own vital statistics could interface with smart homes, and various machines could be used to maintain health and administer drugs.

Autonomous systems of all kinds, from smart homes to vehicles, could be created today. They are not being created because the legal framework has not yet been established to do so, but this is changing quickly. Actually, smart homes, ubiquitous computing, and many other automated systems are being tested today in South Korea. This has been done specifically because of that country's legal structure and culture. The citizens simply don't care that they are test subjects for technologies developed in the United States. They are too busy living life in their virtual worlds to care about such things. From the start, privacy was a non-issue for them. This is exactly the kind of population that is needed to speed up technological progress.

It is mentioned in the British Royal Academy of Engineering's 2009 report on Autonomous Systems that the move toward an automated highway system is being "achieved by degrees." They mention the fact that most people voice opposition to technologies while at the same time using them. This acquiescence constitutes acceptance of their terms of service. In this example may be the biggest problem of all. Our situation has been normalized incrementally, but it is far from normal. We can all sense that something in our world is not right. The burden of this monster weighs heavy on us all. Simply admitting our discomfort has become so difficult that we choose false remedies provided by the very system that made us sick in the first place. We drug ourselves to numb the pain. Quick and easy solutions are what most of us take.

Of course autonomous vehicles will be far safer than any human-driven vehicle ever was. This is not a problem. The issue here is that of enormous government. The Autonomous Systems report mentions that,

> governments will have to play a bigger part in controlling the highways system if autonomous vehicles are to be introduced.[119]

This is a non-issue, if you believe that such a system will be our benefactor. More than anything else, developing converging technology depends on our attitudes. The promise of safety and security must be more important than our own personal burden of freedom. If we are to live longer lives, we will need to focus on safety. That is to say, when life is extended, the threat of a violent death becomes more urgent. No matter what technological developments may come about, it will be the large governmental structure that continues to provide for us, while keeping us safe.

Another point on safety and security concerns the emergence of these technologies themselves. Their power is so great that if they fell into the wrong hands,[120] we would all be in trouble. Who do you think will keep us safe from the ever-present and vaguely defined bad guys? Who will the bad guys be? Keeping us safe from any possible abuse of technology, our loving government will take every precaution to make sure that we are all behaving. It will be in the name of security that we are monitored completely. Of course mere monitoring is nothing compared to the kind of control that is possible to exert over an enhanced population.

Individual empowerment must be mentioned again, this time as it relates to safety and security. One of the best selling points that transhumanists have to promote their cause is that of freedom. Ideally scientific developments would bring about technologies that could truly empower the individual. This is definitely a good idea, a noble concept. What we need to realize is that the entire security apparatus that has been set up is the foundational element that these technologies will be built upon. Their extreme benefits will not be free to use as you wish, there will be extremely strict conditions upon which they will be distributed. We may well be given "desktop nano-

---

[119] Royal Academy of Engineering, *Autonomous Systems*, page 7

[120] The *wrong hands* is a loaded PR term if ever there was one.

assemblers"[121] but we certainly won't be able to use them for our own unique projects. Can't you see how this is going to work? Only officially sanctioned and accepted products will be lawful to build with nano assembler machines (or whatever technology actually ends up existing). No doubt, the large corporations, which are really an arm of government, will be allowed to produce the designs for everything these machines can create. If any person were allowed to freely build whatever they wanted, then it would be deemed a major security issue. Can you understand how our current security overlay fits into this equation? You can forget about the free market system, because it won't be allowed to exist. A megalithic governmental/corporate structure will rule in the name of security. All will be monitored (and owned) by this system, including us!

All the policy papers mention the security risks inherent in these technologies, and we can plainly see where the wind is blowing. A national security state is being built so that a controlling oligarchy may be secure in their dominance over everything. All emerging technologies are well known, because it is the military industrial complex that has developed them! You have to be incredibly oblivious to not understand this fact. The transhuman fundamentalists are well meaning, and they are certainly not wrong in praising advancement. However, in the thrill of their futurist speculation, they have been blinded to what is happening all around them. They don't see that anti-terror legislation has all been aimed at them. It is merely a means by which government can exert complete control over the people. If we keep going forward without questioning where we are at already, then nothing will improve. Forget about the Singularity, because it is not what you think it is. The future is being created with each passing second. Understanding this straightens our perspective. We can't get caught up with future developments until we sort things out here and now.

Tyranny is being built one step at a time. Look at how far we have come already. We don't want to admit that our bodies are viewed as commodities. People were reduced to this status long ago. This is incredibly sad and it should be something that we all take offense to. What we are seeing in the transhumanist project is merely the completion of a giant, long term, technical project. The operation of international business must be streamlined. It is well known that corporations lay off thousands of employees without warning.[122] Such

---

[121] A household nanotech printer that could produce any substance imaginable, including food.

deeds are done without the least remorse. As a legal person, the corporation doesn't share the same feelings that an organic person naturally has. They are beyond such trivial things. They will devastate people's lives merely to cut costs and to operate more efficiently. Is it any wonder that this very same attitude is being applied to human life at a global scale? When you rise to the level of the Divine Man, the macrocosm, you lose track of all of the tiny "cells" existing in your body.

We all need to realize that the devil is in the details. Every reason in the world has been given to substantiate this cold operation, but none of them are worth a dime when it comes down to it. Men and women should not be treated this way. We all need to pull our spines out of the closet and start demanding our own rights, forget about those of machines! We're sick of being machines ourselves.

---

[122] Actual contractors specializing in psychology are brought in to see that violence doesn't break out amongst the displaced workers.

-Chapter Twelve-
Alchemy Throughout the Ages

The Rock of Ages

Alchemy is the process by which one turns base metals into gold. Many people hear this and automatically dismiss the idea altogether. Our modern education system has programmed us to act instantly and without real thought. We have been made to only accept quick and obvious answers to all questions. This is why Alchemy goes completely unnoticed today; however its influence only continues to grow. The exponential rise of Alchemy coincides directly with that of technology. They are one and the same thing. It is quite appropriate that as Alchemy advances, it becomes ever more invisible. This is the essence of its power.

We have been told that turning lead into gold is a silly practice, a scientific impossibility. We are supposed to view such ancient concepts as outdated and ridiculous, the products of mere superstition. The truth of alchemy eludes us, because we are programmed to respond to its exoteric definition, and go no further. To understand the deep meaning behind the veil of words, i.e. *lead* and *gold*, requires that we understand alchemical symbolism. We must choose to go beyond our programmed responses in order to do this.

It is true that alchemy is a science, and a very closely guarded science at that. It is made up of formulae that allow the adept to gain power over the world, to intelligently manipulate matter. This is why it has forever remained encoded through a variety of ciphers. Intelligence agencies have existed for ages, and their Royal Secret has always been closely guarded. True alchemists have always been members of secret sects. Knowledge is power and accordingly the Master Alchemists have forever kept their secrets hidden from the eyes of the profane. To all those worthy of such knowledge, the truth lies hidden in plain sight. The accepted method of attaining higher knowledge of the occult sciences has always been through ritual initiation. This is how one gains entre into the inner circles of the intelligent few.

Fig. 12.1 Symbols, gestures, and numbers abound in alchemical art. Why do certain figures point up or down? What do the subtle differences mean in shapes?

Metaphor is a method by which alchemy is encoded. Elaborate stories, artwork, and pictograms contain formulae. The highest secrets lie encoded within symbols. All mystical practices and esoteric religions utilize this technique. When looking at many ancient Cabbalistic, Rosicrucian, and alchemical pictograms, one gets the distinct feeling that they are looking at technical diagrams. These works of art are actually manuals that employ mystic ciphers. The depth to which such ciphers reach is absolutely amazing. It is not the goal of this book to detail alchemical symbols, but merely point to the fact that there are indeed codes within them. Endless volumes could be produced to decipher the many alchemical riddles of the ages. The greater point to be taken lies in the design itself. We must understand the logical foundation upon which our world has been built. This foundation is in fact a self-replicating pattern. It unfolds into infinite complexity, but that complexity all began with one tiny seed. The Grand Master of alchemists has perpetuated its own will throughout time. We must now become aware of this, and perhaps we may regain much of what we have lost.

Word has it that we are now approaching the end of time, the Eschathon. This is very frightening to most people, and rightfully so. Many have been convinced through deliberate alchemical magic that

Alchemy Throughout the Ages

the end of the world is actually here, and in a way this has sealed their fate, but only if they continue living under the spell of the Grand Master Magus. Absolute ideas create absolute realities, but the amazing truth is that belief in absolutes is a fallacy! The end of the world is nothing more than an idea that has been placed into our minds. If we believe that it is inevitable, then we will make it come true. Events may be forced into existence through attention and forethought. That is to say, we direct our lives according to our beliefs. Whether our beliefs are good, bad, helpful, or harmful does not matter. We will act according to our program. Know this reality, and you will begin to understand the amazing truth of your existence.

We have to understand that our very history has been entangled in an ongoing alchemical project. Our natural, national, and mythical histories have all combined to form a singularity. They have become a powerful force, a catalyst for change, otherwise known as alchemical transmutation. Time represents the motion of the Great Work within space. The end of time would actually bring about a new beginning. On this note let us delve into the symbolic beginning of the Great Work, which has been encoded within the ancient creation myths of the world. Let's go back to the start.

Look at various creation myths and you will see that in the beginning, there was darkness. Out of this darkness arises our mythical identity. The creation of Life marks the beginning of the Great Work. Alchemy is concerned with Mankind/Life itself, and its evolutionary process (mystically referred to as *becoming*). Through man, a perfect spiritual evolution of all life is eventually to be realized. The dark, black Abyss, which is the ubiquitous symbol in these myths, is actually the blank slate of pre-creation.

Take for example the pre-creation myth of the Numina of Greek/Roman mythology. The Numina were known as "those that are above;" they were pure spiritual beings that existed before life itself. They were also known as "the Wills." The power of the will, or will power, is the collective Numina, and this is the highest deity within the Mysteries. The Numina represent a collective, transcendental identity. The multiple Numina are but one powerful force: will power. Will power is believed to be pure spirit, the actual initiating force that brought matter to life in the beginning. This is how the mystery of life is explained within the ancient creation myths. Before life, there was nothing but a dark mass, the Abyss. This was the beginning of the work and it is symbolized in the Caaba stone. One of the original Numina is Saturn, who *is* the Caaba stone. He is the black cube, as well as the lead of alchemy. He is both the beginning and end phases of Alchemical transformation. He is death itself. Death and rebirth are encapsulated in the identity of Saturn. This is why he is known as the patron saint of the alchemists. He is the Master of death and rebirth.

The black Abyss is the primordial darkness that existed before Life. Life is the *wind born egg*. It is the light that arises out of darkness. The darkness is actually the cold dead expanse of the material universe, space. Somehow, some way (of which no one knows for sure) life arose out of this darkness. It is upon this profound mystery that the initial creation myth builds its story. The cold black father is symbolized in Saturn. During the Golden Age in which he ruled, all that existed was dark and cold. Ironically complete darkness was his golden age of immortality, because it was completely devoid of any life whatsoever. You see, where there is no life, there is no death and therefore the Numina existed as immortals. They were pure spirits. Somehow these spirits are believed to have willed life into existence.

Saturn is also known as Chronos. This name points to his identity as the Master of Time. Chronos, or Chronology, is Time itself. This is the form that Father Saturn takes upon the rise of his Sun/son, Jupiter. Jupiter, also known as Zeus, actually represents the physical sun. When he is born, he takes over and becomes the new high Father figure, the Master of Mount Olympus. The mystery of the sun itself involves the fact that it was somehow created by a dark (unseen) father. The sun, which brings life to the abyss is the Light of creation. Light is needed to have life at all. The Sun is the gold of the Alchemists. It is the seventh metal and the highest within the hierarchy of creation. Remember though, that pure spirit is only achieved upon death. Saturn, the dark reaper, the character who holds the sickle, brings this about. He is time; the immortal and unseen force that causes living matter to return to its source. The Prima Materia is pure spiritual matter, it is the Saturnine antimony. Understand this and you will see how the Mystery School operates. Dark matter is revered.

The golden age is the ideal lost estate to be regained by the Alchemists. To literally achieve physical immortality is to complete the Great Work of creation, and become one with the Father himself. This is the godhood to be achieved by the fallen angels. It is the esoteric meaning behind the end of time. When material bodies no longer die, they will be returned to divinity. Time will no longer have relevance as immortality becomes the norm. It is through the mystic science of Alchemy that this high ideal is to be achieved. This is how

Alchemy Throughout the Ages

Man has come to believe that he may actually become God. By perfecting his own nature.

The mystic task of *becoming* involves the art of Alchemy. This art is achieved over time to eventually abolish time itself. Think about what time actually is, and what it has brought us. Our memory of time, of the past, is History itself. History represents our collective memory, and it gives us our identity. Historical myths actually motivate our present actions. To really see into the depths of History, one must understand that it is an allegorical myth in its own right. Our history has been rearranged to such a degree that we have lost the ability to filter truth from fantasy. Taken as a whole, myth and history combine in the art of Alchemy. The Master Alchemists understand that spiritual dominance over time is necessary to guide the dark mass where they want it to go. The darkness of our ignorance is actually part of the prima materia that these alchemists shape according to their divine will. Our very history has become a record of the Alchemical process itself. It is a record of Man's transmutation.

Before official history begins there is *pre*history. The Prehistoric era has unfortunately become a dark abyss of chaos and confusion for we don't have any real knowledge of what went on at such time. All we have is myth, which is often portrayed as scientific fact. If we recognize alchemical symbolism, then some very important aspects of the historical myth will emerge from out of the darkness. We refer to prehistory as the *Stone Age*. If this is not an alchemical metaphor then I don't know what is. You see, the *Stone* Age represents the beginning: the black stone or lead, the Caaba, antimony, the beginning of the alchemical process. Mankind in the Stone Age was the rough ashlar, the Stone to be perfected. So, you see, Man is the Stone of the Stone Age, the same stone that represents the primaterial base that is eventually to be turned into philosophical gold. Throughout time, this stone is transmuted into different types of metals. These metals are actually symbols for the different ages of human history. The Stone Age, the Iron Age, the Bronze Age, all of the metallic ages represent different degrees that the macrocosmic Man has been initiated into throughout time. Also realize that man's discovery of fire is a highly charged metaphor. Fire is symbolic of ingenuity, intelligence, enlightenment, and taken literally, fire is what enabled man to create more and more technologies. During the Stone Age the alchemical furnace of Vulcan was built. This has enabled Man to build more and more creations, as well as build himself in the process. The higher and higher levels of technological achievement that Man has attained throughout the ages correspond with his mystical evolution. Saturn, the Demiurge, or the Ancient One (Time) has remained the initiator, the one who oversees the death and rebirth rituals of the ages. The Alchemical prize lies in the realization of a golden age. This is the new age to come, the Dawn of the New Day. The Light of the New Age shines at the height of creative potential. At this zenith Man will have transcended himself to reach the level of a god. This is what the supposed end of the world is all about; a total transformation of life as we know it. Upon this transformation Gold/light is to shine forth as Man's perfected new body. Also realize that the current *Information Age* is leading up to the attainment of the Philosopher's Stone. Information is immaterial, and is actually a reflection of the Will. Information is a spiritual catalyst for change.

There is yet another great myth of the initial creation which deserves to be mentioned here. It is in fact an elaboration of the pre-creation Abyss story. In this myth the gods "experiment with metals," or in other words carry out Alchemical science projects. These experiments actually represent the Genesis of Man, and the gods are depicted as the original genetic engineers. This particular story begins with a race of men made of pure gold, and ends with a degraded fifth stage race of iron men. We are actually this miserable iron race, that is to say Man in his current form. The immortal golden race of pure spiritual men is in fact the goal to attain through alchemy. Forward progress is made through the transmutation of iron to higher gradations of metal, higher levels of spiritual purity. As a spiritual science, alchemy seeks to destroy the duality of material existence in favor of ultimate unity with god the father. When we see that this particular creation myth depicts the perfect golden race as consisting entirely of men we must realize the implication. Before the duality of material existence, men needed no one to help them procreate. This was the golden estate of immortality symbolized as the primordial Adam. Eve/Pandora brought great sorrow upon the world simply by being born. Her *box* brought forth everything that is wrong with life, save one thing, Hope. By now you should better understand the esoteric agenda of postgenderism. It is all about the destruction and abandonment of all things feminine. The transmutation of humanity and the subsequent abandonment of gender is the way for *Man* to become one with god.

The alchemical creation myths actually reference the imperfection of the natural world itself, of Mother Nature. Her slow evolutionary process not only takes too long, but is also seen as derogatory to Man the god. This is the esoteric meaning behind these creation myths. The original Fall of Man ruined his divinity, and ever since Mother

Nature has only continued to seduce him to fall further into the depths of her womb. Strange? Disturbing? Ridiculous? Even if the answer to all of these questions is a resounding *yes* it does not alter the situation we find ourselves in now. Strange things are afoot, and we continue to remain ignorant. The alchemical process is in fact being revealed in these end times of Revelation, but who among us can see it?

Lucifer the fallen angel is Man, the materialized fallen god that is tasked with finding his way back to paradise. Symbolically, this tells us that the natural form of man's body, his very existence, is actually his greatest adversary. This is why Lucifer is known as the adversary, he represents the prison that is physical existence. Fulfilling the Great Work requires the attainment of a higher spiritual ideal, a whole different form of life. Man must end up destroying himself in the alchemical process. This is why Saturn (death) is the patron saint of the alchemists. The occult sciences are concerned with self-destruction. The natural history of our race has been tied to a mystic allegory of deconstruction. Dionysus, the Natural Man, is also the material body of Man, the green natural expression created by red sunlight. The Sun incarnate on earth is the green man. Dionysus, Osiris, and Lucifer are all different names given to this green man. The emerald tablet of Hermes is yet another green representation of living intellect. The Lapis Exilis, the green stone of Lucifer is as well. Through proper use of his greatest gift, his mind, Man is to attain true spiritual perfection. Only by perfecting the Philosopher's Stone (the mind) may this unfortunate fallen god return to divine unity. This is what the Son of God, Osiris,[123] must do to return to his Father.

The father of Osiris is Saturn. Osiris is the sun, which is also the son of Chronos himself. He is Life, Light, and the way to salvation. He is the dying god, who constantly labors toward his own perfection through Enlightenment. His task is to recreate his Father's golden age. Why do you think the plan has long been to establish an age of reason? The golden age is the light of Rhea's son, Osiris. Rhea is the wife/sister to Saturn and the mother of Osiris. Rhea's son is *reason*. Osiris represents perfected reason within the mind of the Enlightened man. Perfected logic incarnated in the material realm is believed to be the only way to achieve immortality. This is the great natural gift given to fallen humanity. It is the fire of Prometheus. The fire of the sun is made manifest within green nature. This is the Fire of the mind. It is the highest expression of divinity within this world. Science is

---

[123] Osiris is also depicted as the Green Man.

the byproduct of this fire, and the means by which pure spirit may be imbued into matter.

Only through Art, otherwise known as Alchemy, can Nature be conquered. Throughout history the Osirian god men have conquered the natural world and all those truly natural people standing in the way of their Great Work. They are completely dedicated to fulfilling the promise of their Father, and they shall never be satisfied until they obtain his power. These deranged characters are completely intent on fulfilling an ultimate death and rebirth ritual, which has as its aim the end of Time itself. This is the high esoteric truth that lies behind the myth[124] of Armageddon.

The Builder

Ages are symbolized in different metals, different phases of the Alchemical work. Revolutions speed up the Alchemical Great Work of ages. Revolution, or rotation, causes the Wheel to spin. The wheel is the zodiac, which is actually a great timepiece. So, as revolution is sparked, great catalytic change occurs and the art of alchemy is carried out quicker. Time effectively speeds up, not in the literal sense but because of more rapid growth. This happens sociologically, scientifically, economically, etc. Alchemy steadily builds up man and as this occurs, its very own processes and effects speed up. The sciences build upon themselves. Technology builds as all of its divisions converge. This is exactly what the promoters of the Singularity are talking about. The exponential rise of technology is actually the perfection of the alchemical process. When terminal velocity is reached and the Singularity occurs, then Time will have effectively ended. This is the end of time that we have heard so much about from the legends of the Mayans to those of the Mormons. The last days of this age are here. The Alchemists are now preparing for a grand ritual that will initiate the entire human race into the final golden phase of the Great Work. The base metals (man) will finally be turned into gold (the new man). The New Man is a wonderful spiritual being but he is also a product of science. He transcends his human limitations to become completely posthuman.

Before time ends, it speeds up to an extreme degree. Degrees are units that measure heat. When the parts of any system speed up, they get hot. Solid substances have slow moving particles, but as these particles heat up and go faster, the solid transforms to liquid and then

---

[124] Please notice that I say *myth*, and not truth. Armageddon is merely an idea.

Alchemy Throughout the Ages 259

to gas. Gas rises because it has the fastest moving particles. This is a scientific property of matter. The various parts of an internal combustion engine get incredibly hot when they are in motion, and thus the need for motor oil and coolant to keep them from melting. Alchemy is a science, an occult science that has these same properties, although they are for the most part symbolic. It uses heat to melt metals and subsequently purify them through transmutation to higher degrees of purity. This is the nature of heat and of fire. The fire of creation burns things up, supposedly up toward heaven. Mystic fire, that element brought to Man by Prometheus, is finally used to its fullest extent as it burns the body of Man at the altar. This is done to initiate his new higher level of consciousness. To become reborn, born of fire like the Phoenix, Man must go through this trial by fire. The Old Man must burn so that the New Man may rise. This symbolism is ancient but you may just recognize it from the *Burning Man* festival that takes place every year in the Nevada desert.

The end of the world is not what you think it is. Within this scientific system the end is only the beginning. We have been thrown into a cyclical process that has patterned itself on the laws of nature. This artificial process claims to be a perfected version of Natural (r)evolution. "By art much better than by their Nature;" this is the line alchemists' use. Achieving the highest spiritual goals requires going beyond Nature to attain something more pure. Doesn't this sound wonderful? Mystic allegory is meant to inspire our curiosity so that we continue onward into the unknown. However, that which remains unknown to us is well known to those who hold the key to the intelligent design, the ones who can read the legend. The priesthood has always known where they were going because their plan follows a specific design. The occult inspiration for science is not to be seen by us. Science is meant to blend into everyday life. This is a problem. We are being led to partial understandings so that we unconsciously fulfill the will of someone or something else.

> I am Osiris the King - I am the eldest son of Saturn: I was born of the brilliant and magnificent egg, and my substance is of the same nature as that which composes light.
>
> -inscription from an ancient Egyptian pillar

Let us return once again to Osiris, the sun god of Egypt. He was the sun incarnated within earth as man, the green man. Scientifically speaking, it is sunlight that actually energizes life on our planet. Its fire speeds up dead matter and animates it to the status of life. Man's existence is secondary to that of the sun, and it is also seen as an extension of the sun's very essence. Living light is Man. The green Natural Man is symbolized in Osiris. So Osiris was the sun as man. This ancient motif has never ceased to influence our religious myths. Osiris was to ancient Egypt what Jesus is to modern Christianity. Now, it is obvious that most Christians don't view Jesus Christ as the material incarnation of the sun itself; however they do believe that he is "the truth, the light, and the way." They also worship him as god's literal son. To them, Jesus Christ was an actual man who lived two thousand years ago, but he was also the "only begotten son" of an unseen Father. This son of an unseen god was somehow sent down to earth and incarnated as a man. This literal man is the physical idol that Jesus becomes. He is transformed from a spiritual idea to a physical person, a historical figure. In ancient Egypt, the people truly believed that the spirit of Osiris was present in the sacrificial Apis bull. Belief is a powerful thing but it is also a major problem. The masses have consistently been led to idol worship. They worship objects while believing they are piously worshipping the true creator.

The simple fact of the matter is that Jesus is the same old sun god. He is the hero character that has been created to inspire us. Heroes have followed the same model throughout time, and how is it that the process of alchemy could continue throughout the ages? Success in the Work has occurred due to continual adherence to a standard: a gold standard when it comes to religion. Sons of god appear in every age because they are the same deity brought back to life in the minds of the people. A belief system perpetuated throughout time continues its forward progress. The sun rises again and again. History has been one unbroken chain of people following the Light of their god. Little did they know that by worshipping this light they were merely worshipping themselves! Jesus takes the form of a man for this reason. Self worship, whether it is done consciously or unconsciously, ends in the same way. It actually turns myth into reality. Powerful belief becomes a self-fulfilling prophecy. Myth and history have intertwined like the twisting Caduceus of Mercury. Man is now becoming god because that was the philosophical gold that the myth always encapsulated. Any true creator is perfectly fine with this entire show, because it is just. We are all getting what we want. In this time of revelation, it behooves us all to sincerely confront ourselves so that we may know what we truly want. Do we want to play our part in placing the capstone upon the Alchemical Great Work, or do we want something else?

As the formula has remained constant throughout the ages, mankind has been built up to greater and greater heights. The Tower

Alchemy Throughout the Ages

of Babel is now approaching the seventh sphere, that of gold. The golden new age is now emerging because we have all been complicit in building this tower. Due to the sheer magnitude of this project, individuals have all too often been forced into compliance with its design, its standards. This is the power of scientific progress. It completely crushes the individual, or does it? Regaining freewill is a task that only the individual can undertake and so you must find that answer for yourself. What can be understood plainly is the Osiris myth. It takes some effort, but once we discover the *Lost Key*, the legend to the map, we can decode history itself. The Mystery unravels, at least enough for us to see what we have been perpetuating for so long. Out of necessity, lies have posed as truth, but what is the real truth? There can be no doubt that we (the masses of humanity) have unconsciously been slaving away for eons! All so we may fulfill the so-called divine will of the Builder himself. What kind of an end comes from such beginnings? An eternally unconscious slave when turned to gold can only become a perfect slave. Is this our fate? Are we unconscious, or merely asleep at the wheel?

> There was a time, the Golden Age, when truth and wisdom ruled the earth, and this state of being by the firm kindly hand of the enlightened sage. This was the divine dynasty of the mythological priest-kings who were qualified to govern humanity by virtue not only temporal but by divine attributes.
>
> -Manly P. Hall, Freemasonry of the Ancient Egyptians

Rhea's Son, Osiris, the King of Egypt, goes throughout the world to spread his civilization through the "persuasion of reason." This is the method by which he builds. Remember how military PSY-OPS work? Through reason society expands as it is standardized. In this way democratic society has been built up in our time. Osiris is the Builder. The Masonic legend of Hiram Abiff (also known as the Builder) is actually another interpretation of this story. When you decode the symbolism of the Mysteries you will find that many seemingly separate characters represent but One idea. This is the Singularity; the great oneness that stretches throughout the ages and limits itself to no creed for it creates them all. Hiram, or Chiram (Chi being the number 10, X, the tetractys, etc.) is the Builder, the one whose task it is to build the Temple of Solomon. More appropriately his task is to build and rebuild the temple. Not only does the temple symbolize human society, but also the human body itself. The body is a temple after all, and all famous temples and cities have actually been modeled on the geometry of the human body. When the goal is to rebuild the Temple of Solomon, it is to complete the Alchemical process of transmuting the base metals of the human body into gold. This is how Osiris is to complete his project of building so that he may then return to his Father.

The Masons refer to themselves as the builders. They build their own lives upon the template of the dying god himself. They are Osiris and that is why they believe themselves to be the Light. Remember that Osiris is actually the physical embodiment of light. That is what the Enlightened Man is. The intelligent, Enlightened Man gains the ability to build greater temples through the use of his mind. The mind is his natural tool that allows him to perfect his own nature. Natural evolution culminates in human consciousness, which becomes self-aware of its own imperfection. Jealousy of the true creator god (whatever it may be) spurs the Fallen One to prove his worthiness of true godhood. The Builder turns his consciousness toward the task of self-improvement: mind over matter. Through the body, the spiritual will finds an outlet to expand itself to greater grades or levels of purity. The dying god actually ends up killing himself so that he may return to his father, Saturn. He wants to become immortal, just like daddy. To think about the body as Man and the so-called spirit or will as some unknown and intangible thing is very disturbing, but it also raises some very valid questions. Could it be that the builder, as a collective, is a powerful entity with no regard for individual human life? Don't rule out the possibility.

Ahriman, the force of evil in the Persian mystic tradition (Zoroastrianism) is an interesting character. All mystic evil is interesting to decode because at the end of the day, the high priests believe themselves to be beyond both good and evil. Rearrange the letters of Ahriman and you get Hiram-an. This could also be read as Hiram Man, another code for Man as the fallen god of the Mysteries, Man the Builder. The only evil these builders see is the ignorance that holds them back from becoming perfected gods. Limitation is a big problem, especially limited intellect. Without the right know-how, the temple cannot be built at all. This is why intellect must be expanded by any means necessary. Forget good and evil, these are both irrelevant in the eyes of the Great Architect of the Universe. The only thing that does matter is that intelligence expands within the physical world. Perfected Logic and Reason will rebuild the temple in whatever way possible. In Manley P. Hall's *The Secret Teachings of All Ages* it says,

Alchemy Throughout the Ages

Reason shall divide the Light from the darkness and establish Truth in the midst of the waters... Light and the fire which rise are the divine man, ascending in the path of the word, and that which fails to ascend is the mortal man, which may not partake of immortality. Learn deeply of the mind and its mystery, for therein lies the secret of immortality.[125]

It is clear that the builder is man but the greater spiritual truth that underlies and goes beyond all of this, will remain elusive. That is how it must be, for truly spiritual things are mysterious. From a certain point of view, this represents a major problem. The Gnostic, the man who needs to know everything, will end up defining his god so that he may later become that same god. A perceived conflict between himself and his natural existence brings about a need to be better. Attaining satisfaction by realizing godhood is what alchemy is all about. It is the gold that the philosopher seeks. This entire process is spurred on by conflict. Everything has been explained in terms of conflict and not cooperation. What we must realize is this particular point of view has had profound influence over our lives. Much human misery is actually a byproduct of this mindset. A very troubling realization comes when contemplating the truth of our situation. How can any satisfaction ever be derived from a process that has always been at odds with nature? How can the Builder ever be appeased? What kind of god will emerge out of the Alchemical crucible? These are important questions to be considered.

Any sort of truly natural existence has been systematically eradicated from the face of this earth. Even the memory of such a life has had to be destroyed from our consciousness so that the alchemical process of rebuilding could proceed smoothly, or at least as smooth as such a violent thing could ever hope to be. Creation, and Man's rise to godhood, requires massive destruction. Death and rebirth are inherent in the Alchemical method. The end times are not going to be fun if we choose to see them through according to their prevailing design. They will actually become the horrors that we all expect. Our holy books and prophets will lead our minds to their predetermined ends. Complete and utter annihilation, at least of what now exists, will occur. This is how the Phoenix rises and brings its light to the new age. We all have some hard work to do right now. The hardest thing is evaluating ourselves and what we actually want. Do we want to be reborn? Do we care to know what it means to be reborn in the first place? Do we want to worship ourselves, or do we find it more appropriate to respect our creator by leaving its mystery intact? What

---

[125] Hall, *Secret Teachings*, page 99

would we be capable of doing if we stopped believing the lies of false gods, those who have made us *believe* great lies?

Solve Et Coagula

Human history has systematically been swallowed up by the alchemical myth. This has occurred to such a degree that natural, social, and religious histories have all become one large myth. Constructed from the material of this myth is a foundation. This foundation is what we proceed to build our lives upon. As a consequence we actually perpetuate and live a myth. Sadly, the actual truth of history has never really mattered. It is continually shattered by myth. This happens time and time again, like clockwork.

When we don't properly acknowledge the mistakes and injustices of our past, then it only follows that further injustice will continue. If we don't properly understand our history, then we will base our present decisions on erroneous information. This is how our problems perpetuate themselves. They arise out of incomplete and incorrect ideas. This is a vicious cycle indeed. It transforms life itself into a struggle instead of a joy. Struggle is actually the fire that fuels the depraved system surrounding us now. Everything is one big problem and this is actually a good thing for demented control freaks who wish to lord over us. They require a degraded and weak population in order to gain the false strength they crave. You see, we are only weak because we have been turned upon ourselves. This makes us malleable. We have become just like metals that are easily melted down. Alchemy has melted and recast us into a variety of machine parts. In this interdependent world of progress, everyone has a role to fulfill: a place for everything, and everything in its place. Humans have been degraded; they have been reduced to the status of mere things.

The dialectic process involved in Alchemy requires first and foremost that a problem be created. This said problem is presented in such a way that a solution is demanded. That solution is what causes more change/progress to occur in the real world. The solution is a catalyst for increased building. Progressive movements work in this way; they always press onward. *Solution* itself is a very telling word. Besides being an answer, it can also be a liquid that has the power to

Fig. 12.2 Baphomet points his right hand up toward heaven from which his divine solution comes. Notice that this arm says *solve*. *Coagula* is written on the left arm symbolizing the lower material coagulation that is made from out of the higher spiritual solution. Notice also that the Hermetic Caduceus rises from the genital area. This cryptic piece of art is all about the godly power of creation realized within the material world. It depicts enlightened regeneration. The dialectic as well as its superior, the trinity, are both seen. The high trinity is symbolized by the fire burning bright in the center and above all else. Right/left, up/down, male/female. Baphomet has it all.

Fig. 12.3 The Green Lion (The natural man) devours the Red Lion (the sun).

dis-solve. A solution dissolves solid substances. This is yet another important aspect of the Great Work of alchemy. You must first dissolve the base metal so that it may coagulate into its new form. This is the philosophical process known as solve et coagula, and it is how the alchemist solves his prefabricated problems. It applies to the dialectic mind manipulation technique, which has been described as problem, reaction, solution. Out of the dialectic comes the divine trinity, which rises out of the waters of the solution. The Alchemists hold their universal solvent near and dear.

The Universal Solvent of the alchemists is the Green Lion, the Philosopher's Vitriol. Symbolically this connects with Mercury, and Mercury, Hermes, or Thoth represents the mind. He is the messenger who flies up and down from heaven to earth on winged sandals (he also has a winged hat). He is the one who brings fire down to earth, intelligence to matter. Mercury, or the Green Lion, is actually the mind/nervous system of the Alchemist. By connecting with the divine mind of god above, the Alchemist gains the ability to shape the material world below. Again, the color green and the power of the Natural Man are seen in this allegory. As the entire nervous system, Mercury travels up and down the spine. The spinal nerves actually twist like the dual snakes of Hermes' Caduceus. Thoth (pronounced thought) goes from heaven (the brain) all the way down to the material regenerative system (the genitals). This allows him to procreate wisely. His godlike intellect allows him to seed the material world in a predictive manner. Not only is this seed symbolic of divinely inspired ideas, but also literal semen, the seed of creation that regenerates a man's body. The Alchemist is always sure to sow each seed in the proper season.

Those ancient and familiar myths of creation all speak of a king who marries and begets children with his sister. In one sense, the king is heaven and the sister is earth. Sis, or Isis, is the goddess of ten thousand names, for all world myths speak of but one goddess. Mother Earth is Isis. Material existence, Mater, or water are all feminine symbols. They are all Isis. The waters of life, otherwise known as blood, sustain physical existence. These waters are present in the lunar menstrual cycle of the female. This bloodletting happens so that new life may be created. Also note that the red blood of man is the wine of Dionysus, the actual solar power of the sun. Dionysus is the god of the vine. Vine is an anagram for vein. Veins/ vines furnish the wine that is the blood. Blood is the soul of Bacchus, his living fire. When this red liquid is purified (through proper use of the mind) it changes symbolic color from red to green. The Red Lion becomes the Green Lion. Hermetics is all about using the mind to intelligently regenerate the body. This is an allegory for selective breeding. This practice is perhaps epitomized in royal families who literally intermarry in the same fashion as the gods. The mythic gods were simply built upon the model of Man, the Man in charge. This is the same man that we all hope to stick it to someday. He may just let you, if you can furnish the appropriate birth records to prove your noble descent.

What we are talking about here, although it may not be obvious, is the actual attainment of the Holy Grail. The cup of everlasting life contains the blood of Man. Blood is the life-sustaining force that keeps a man alive and it is also carries his unique genetic code, his DNA. It is through the purification and preservation of this liquid that Man regenerates time and time again. It is how he achieves literal reincarnation, which is everlasting life. A *pure* bloodline is one that has been methodically maintained over a long period of time and in this way has immortalized DNA itself. If you look at the glyph symbol of Mercury, you will notice that it looks like a chalice. That is because it too references the cup that holds the holy water of everlasting life. The material soul of the sun itself is captured in the cup of Mercury; this is the purified blood of the enlightened ones. The Philosopher's Stone is the mind, the emerald tablet of Hermes. Through this mechanism, Man learns different methods by which he may attain the grail. There is more than one way to achieve the Philosopher's Stone. Different paths can be taken.

In alchemy the process of regeneration/reincarnation can be fulfilled through either the wet or dry paths. Because water is feminine, the wet path represents a perfection of the natural process of regeneration. Throughout the cycles of life and death, Man procreates through typical breeding methods. If done properly, he regenerates; he lives on in his progeny. If done foolishly, without forethought, he degrades. The wet path involves calculated mixing and perfecting of bloodlines. In the high mystic tradition this is done through selective breeding. If you want your children to remain uniform, to be just like you, then you have to pick the proper mate. This is how the incestuous practices of royal breeding arose. The priesthood actually oversaw the entire process. They approved marriages between high profile royal families. This was done for varied strategic reasons. Not only could controlled breeding perpetuate a certain type, it could also be politically beneficial. By joining two powerful families, geopolitical power moves were made. The wet path of alchemy, overseen by the Hermetic priesthood, was believed to be the height of natural evolution. Through enlightened spirit, the chaotic lusts were tamed, and intelligent design took place. This was how the Dionysian god men legitimated their wise rule over the material world.

Fire and earth are the two elements that symbolize the dry path of regeneration. They are two opposing poles of the zodiac that represent South and North, Summer and Winter. Bringing fire to earth is that same old Promethean concept. The idea behind the dry path is to imbue fire (intelligence) directly into the cold dry earth. This method involves the use of material science to bring about an artificial form of evolution. Through the intense heat of fire, this process is meant to be fast. Rapid evolution that far exceeds the speed of Mother Nature will commence. Genetic engineering is the ultimate way to purify the blood. By preserving his blood in the cold of winter, the dead man may literally be brought back to life. Winter is symbolic of death but the dry path has the potential to cure such ills. Reincarnation could be achieved in its most literal sense by using DNA to construct new life. This artificial process is the highest Alchemy; it is the goal of the priesthood. They want total efficiency in reincarnation and they know how to get it. On the branching road, the Y, the dry path represents the right way to go about turning the base metals into gold. It isn't as messy and uncivilized as the wet path. It is completely calculated and predictive, a perfect science.

The Hermetic solvent of the Philosophers is the solution to their grand problem. It dissolves all impurities to turn the base metals into gold. Gold is a symbol of wealth. By genetically engineering an entire population, the Master Alchemist would become unfathomably wealthy. The former base metals, or base model slaves, would be upgraded to golden, top of the line, high performance machines. This is all the Master has ever really wanted; servants who completely understand their place in the Universe without ever complaining about it. In fact, the new golden slave could be efficiently programmed to love his status.

The self-proclaimed gods have attained their status through their Hermetic method. They have conquered their lowly animal instincts to rise above as natural men. Everyone knows Man is at the top of the food chain in this world, but not everyone realizes the esoteric meaning of the word Man. The true Man is an evolved spiritual being of light. He has attained godlike intellect and has mastered the beast within. As a true man, he rises above the animal kingdom to conquer the beasts below him. The number of the beast, 666, is actually a symbol for the lower animal nature within man. In numerology it reduces to the number 9, which is the occult number of Man as well as that of Alchemy. Alchemy is all about conquering the beast, taming Nature, and controlling the world through enlightened reason.

The four lower elements are actually subordinate to the high trinity of the mind. The four elements are the square (which is also symbolized in the block, and the physical foundation). The three aspects of the trinity create the artwork that is placed on top of the base. This is how the base level, unreasoning, unthinking elements of Nature are conquered. Dionysius Andreas Freher has written about the "world spirit" created by the four elements. This lower spirit affects man accordingly,

> it forms him as such an animal in the outward life-property, for the 'Spiritus' of the outer world of the elements cannot give other than an animal

Naturally formed animals equate to the masses and the naturally formed Man is the Philosopher. The natural byproduct of this philosophy/religion is that you find a small minority, a Philosophic Elect, lording over a population of servile beasts of burden. This amounts to the *Animal Farm* of George Orwell. That book was a powerful allegory that probed our social construct as well as the minds of the various animals living within it. We are being farmed. This is being done logically, scientifically, and methodically. The Method of the Alchemist is used to perpetuate the seasons and see that the proper harvest is reaped. To keep the masses motivated, the empty promise of the windmill is perpetually given to them.

We are viewed as animals and proclaimed to be weak. In the eyes of the priesthood, the unwashed masses are confounded because they allow the regenerative system to obtain control over the mind. If they were enlightened the reverse would be true. By not intelligently controlling and subduing lust, we are no better than the animals. Instead of taming the 12 animals of the zodiac, we allow them to overtake its 13th aspect, which is the sun itself. The sun is the mind and it has the power to actually profit from the 12 animal signs. These are the houses that the sun travels through on its journey across the sky. Because our animal sign becomes our identity (exoteric astrology works this way), we in turn drop the torch of enlightenment. We go from the status of human to that of animal. This is where the term degenerate derives. The masses are viewed as degenerate creatures, not men and women.

In another sense, the animal masses are actually viewed as domesticated creatures, bred for a specific task: slavery. They are mindless and passive workers who never rebel or complain. Their place in the world is to work, not to think. The enlightened ones substantiate this aspect of their religion by observing a complete lack of resistance or even concern from the multitude. Hell is perpetuated by indifference. The millstone keeps grinding.

Dissolution is an important part of the alchemical process. It is what breaks down the solid metal so that it may later coagulate into a new form. *Disillusion*, chaos and confusion; these all represent the dissolution phase of alchemy for they break down the old. When people no longer know what to believe, they break down mentally. This sends them desperately looking for the safety and security offered by a belief system. In desperation just about any answer to their problem will suffice. Anyone in a position to offer such answers or solutions will be able to obtain great power. Don't think for a moment that this potential power is ignored. The dissolution phase is chaotic by nature and this is actually a good thing for the initiator. It is he who knows where the candidate is being directed; he knows the final outcome for he designed it. Solutions require a problem to even exist in the first place. Without a perceived problem they are worthless. This is why our current Age of Transitions consists of one

Alchemy Throughout the Ages

problem after another. We are all being broken down, dissolved. This is how all of the worldviews, religions, traditions, and standards of the passing age are dissolving alongside us.

The coagulation phase comes after the metal has been fully dissolved. We, the human population of earth, are to be transmuted into gold. Coagulation involves bringing us together as a liquid hot substance so that we may be poured into an enormous mold. We are then to be hardened into a macrocosmic unity. Society and its inhabitants are all to be one so that peace may reign on earth. This is actually what we will believe is happening, if we allow ourselves to break down in the first place. Peace will literally take over our minds. However, this peace will merely be passive compliance. Slavery will be labeled freedom and we will believe it. Belief is a powerful thing. This is a time-tested fact. When you throw high-tech human enhancement technologies into this mix, things get absolutely horrific. The hive mind would govern the posthuman body politic. A new standard for a new age could be scientifically designed and maintained precisely. Any semblance of freewill and individuality would be crushed. This is the idea and you must realize these things now. Chaos is to be continually conjured by the masters of the Black Art until the coagulation phase of their Great Work is come. The fate of the entire human species now rests with you. Your very consciousness is the only thing that can stop this from happening. Is this too much responsibility for you? You or I don't have the luxury to default on this duty.

The entire monstrosity enveloping us, consuming our bodies, minds, and spirits is completely rationalized and explained away through logic and reason. This is how its evil is disguised. When the very concept of evil disappears, then so do our cares. We cannot believe in something that doesn't exist. The amount of work that has gone into the Great Work of ages is staggering and cannot properly be understood by anyone who exists outside the walls of its military fortress. We cannot know the complete depth to which this Hell descends, but from our vantage point we can see more than enough to understand that we've been duped. If we choose to do so, we can smash the lies that have been thrown in front of us since birth. We can prove that we are more than just animals, slaves, and degenerates. All of the human shields thrown out to take the heat for their Unknown Superiors will be tossed aside as truly educated men and women march forward. This is something that the alchemists cannot have. Their great power is actually great weakness, and as such it requires that we all be weak in order for it to exist at all. Their divine will is a fraud, their inner religion is an absolute disgrace, and their very existence is a cancer upon humanity. It is now time to find a cure for this cancer, and we don't need to donate one red cent to any foundation in order to do so. All we need is to self-actualize and empower each other. Strong individuals instinctively work together to help each other. They know false idols when they see them, and they know what to do with them. Ask yourself this one question, but before you answer it, think long and hard to make sure you know how to answer it: Do you feel comfortable worshipping a demigod, a man, a materialized intelligence, or would you prefer acknowledging the true higher creator in all its mystery?

We have the ability to do anything we want to do. Freewill is real, and it belongs to you. This was always your divine gift, and in truth no one can keep it from you. The secret behind technological enhancement is that it is actually coming about to cut you off from freewill. It is now your responsibility to reclaim your true power, and dispose of its false alternative.

-Conclusion-

What are we really? The answer to this question is not as simple as most of us would like it to be. We are often discouraged by the fact that absolute answers are not forthcoming, at least not to eternal questions like this. In the absence of a literal deity standing before us doling out answers to all of life's eternal questions, clever human beings have stepped in to take the place of true divinity. These priests understand our frustrations and actually use them as leverage to increase their own worldly power. Our impatience leads us to the convenient conclusions placed before us by an unseen priesthood. These conclusions eventually get taken for granted and become the foundation that we build our entire world upon.

The terribly wonderful reality underlying our existence is that of unlimited potential. That is to say, any idea we entertain can be made to manifest in time. *Anything* may be made into reality if the effort is put forth. Whatever we want to do in life can be done. We do have freedom but only if we demand it. A severe lack in demand for freedom is our real problem. This goes unnoticed due to our skewed understanding of what freedom is. We have actually been convinced that certain idols such as money, flags, and cars epitomize freedom. By narrowing our vision, a grand illusion has been pulled off. Broad understanding is not something that the priests of power want us to acquire. That would endanger their petty power trip, their self-proclaimed godhood. This is why our minds have been divided. As the world becomes more complicated we each become more intensely focused on one small field of study. We have become specialized machines programmed to fulfill very specific goals. Is it any wonder that we worship high performance machinery?

Logistics is all about getting things where they need to go. Human beings were reduced to the status of things long ago and our present situation is nothing more than a logistical success story. The logical conclusion to an age-old foundation is working itself out and we are none the wiser. Actually, we are wiser, and we do instinctively sense extreme danger all around us. By acting according to *reason*, we decide to brush inconvenient feelings aside. Science has actually given us proclaimed medical cures to help us do just this. We quickly swallow down pharmaceuticals to suppress our falsely labeled problem, our emotions. The true identity of this problem is our survival instinct. Our senses have been dulled. We have been trained to look at the world practically and empirically, to analyze problems according to logic. This sounds good, but the real question we need to ask ourselves is whose logical model are we using? Is it possible that large portions of our mind have been tuned out? Are we now debilitating portions of our human brain to make way for an entirely new model?

It is foolish to believe that technological enhancement will somehow bridge the gap that we have created in our own minds. This chasm in human potential has been deliberately produced and maintained to obtain specific results. Technology itself is merely a product of our minds and as such, it too has unlimited potential. It could be used to do anything. Many people realize this and confuse such unlimited potential with the actual reality of scientific development that is going on all around us. Yes, technological advancement is amazing, but where is it going? What higher purpose does it serve? Who is guiding this development and who will actually possess legal ownership over the end product? Always remember the practicalities involved in this venture. Use true logic to decode the system.

Transhuman enhancement would improve our ability to do what we are already doing. It would increase the power and potential of our current social system. An artificial environment has been built all around us. This has caused human beings much trouble because it is far removed from the natural environment in which we originally existed. Instead of altering our self-created and artificial environment, we instead turn on ourselves. *We* become the problem. It couldn't be our scientific progress that is the problem, oh no, we are at fault biologically. Because the world is so much better than it was in our distant past, progress cannot possibly be derailed. The thought of going backward or of simply stagnating is bothersome, and so we press on. Human beings themselves are now the focus of scientific transformation. Genetically modified posthumans would be altered to fit comfortably within the structure of this artificial system. By merging biology and artifice, the system itself would be upgraded. Before killing off the old man, we are first being convinced that he is inherently evil. Could this entire process be sadomasochistic? Complete self-destruction is being offered up as the solution to ourselves.

We are living in a time of rapid change, of dissolution. Newt Gingrich called this the "Age of Transitions," and Zbigniew

Conclusion

Brzezinski titled his book about technetronic warfare *Between Two Ages*. We are also hearing fundamentalists of many creeds and from all walks of life proclaim that we are living in the end times. Armageddon is on everyone's mind. Chaos and confusion are rampant and yet there is definite unity within the seemingly diffused multitude. The one thing everyone can agree on is that things cannot simply continue as they are now. Major change must occur. This is something everyone is in tune with. A large group of tuned instruments comes together to form an orchestra. Orchestration is organization. A conductor sets the time, and leads the orchestra forward. Illumined minds of genius write symphonies in order to generate a specific response from the audience. Mozart can really light up *The Flute*.

The time in which we are living is a transitory moment of great change. As it stands, we are now headed toward a complete alchemical transmutation of the entire planet. What does this mean exactly? That is impossible to say for sure. The Singularity concept may simply be a distraction. Its high idealism may be a front to drum up support for artificial intelligence, genetic engineering, surveillance, and global government. The transhuman vision is still an idea and its potential remains unlimited. This is the power of alchemy. Powerful alchemists have continually misused and abused the entire planet for their own gain. The evidence of this surrounds us. Unchecked power has run rampant and will continue to do so regardless of our own personal opinions of the world. We need to begin questioning the source of our seemingly independent opinions. Many of us have been convinced that our artificial system has been created to aid us. In a strange twist of logic, this is true. The help that is being offered is that of total destruction. Death is the door to ultimate rebirth.

A perceived need for something better makes us forget what we already have. The real truth is that weakness is a fraud. We actually uphold this lie and limit ourselves. By believing that we must live our lives reasonably, according to very specific directions, we forfeit our true eternal gift. We fall into the intelligent design of the Master Alchemists. Weakness is actually encouraged and exploited. A small minority profit from the mental instability and weakness of the multitude, because it is this that prevents them from attaining profound self-realization. A population of strong independent people who naturally help and look after one another will resist tyrannical government. It is no coincidence that a totalitarian world government is now rising to power, while most of us remain completely oblivious to it. All throughout human history we have seen the rise and fall of empires, of rulers, of tyrants. It has been the hope of ages that one day these despotic creatures would be banished from the earth. We are now entering a time in which this hope is being exploited. It may be possible to convince the world that all the strongmen have been defeated. The issue here is that of perception. Just because we believe that tyranny has been conquered does not make it so. It seems to me that we are steadily being led to a point where we will be physically unable to recognize any pain. By blocking out all of our discomfort, a tyrant could take advantage of us in any way imaginable, and we would not care in the slightest.

Understanding the occult side of world events enables one to actually know what is happening. By reading between the lines, one can see the true story encoded within the exoteric story. A veil of illusion has been constructed and distributed for mass consumption. Mere lies are nothing compared to the power of complete distortion. This is the power of the magician, the Master who wears a top hat and white gloves. He knows how to divert your attention and distort your perceptions. Never doubt the power of magic, of alchemy, of hypnotism. There is a scientific method involved in all of these disciplines. As the mind is further decoded with science, it only follows that the art of alchemy will also improve.

Make no mistake, alchemy is completely real, but what is it? It is nothing more than the art of change; making your will manifest in the material realm. The electrochemical processes of the brain cause the chemical reorganization of matter. Actual electrical currents in the brain are akin to the lightning bolts of Zeus. It is this father figure who lords over the world from his high position on Mount Olympus. His lightning bolts strike down all those who oppose him. It is through the power of knowledge that he rules over lowly mortals. They are the fools who are ignorant of his godly power. It is the foolishness of lower beings that places this particular god in his position. If you look at many of the ancient myths you will see recurring themes of violence. A rigid power structure is maintained and any who dare defy it are quickly put in their place. This is an allegory for the human hierarchy of labor. Everyone has his place and anyone who breaks the rules is quickly taken care of.

Science is concerned with studying nature so that it may later be manipulated. Ultimately, its goal is to improve upon nature by completely rebuilding it. Nature includes ourselves and so we can see that we are to be altered along with the entire planet. The human mind has long been viewed as the most sacred creation in the natural

Conclusion

world. By multiplying the power of this divine instrument, everything existing below it is altered. The idea that the natural world is for the most part useless, or dead, stems from the ancient mystery tradition. By bringing dead substance to life, it becomes useful. It evolves, imbuing intelligence into nature, spirit into matter, and waking the entire universe to conscious awareness. This is the highest esoteric ideal within the mystery tradition. It is the way that man himself may evolve to a higher degree of godhood. The transhumanists often speak to this desire. Nick Bostrom's paper, *Are You Living in a Computer Simulation,* posits that by using powerful technology, posthumans could,

> convert planets and other astronomical resources into enormously powerful computers.

Entire simulated worlds could be created from this intelligent material. Of course the creators of these simulations would maintain complete control,

> The posthumans running a simulation are like gods...
>
> -Nick Bostrom, *Are You Living in a Computer Simulation?*, 2001

Godhood is the prize to be obtained. This divine estate has already been achieved philosophically by the Master Alchemists. Now they seek to amplify their powers. Whether or not these "gods" will obtain complete control over the universe remains to be seen. What is certain is that their ancient power has been maintained through the creation of virtual realities. By creating powerful beliefs in the minds of their subjects, the priests have effectively dictated the direction of human existence. Dictators always rise to power through the clever use of mass mind-control techniques. The very word propaganda is derived from a branch of the Catholic Church that was created for missionary work. The work of the church has always been to unify the minds of the masses. In the new age, science shall rise up and claim this power. It will be the focus of worship for the multitude. Scientific miracles will be achieved, all of which shall be a testament to the wisdom of the established system. Nothing will be permitted to disturb the peace of this immaculate deception.

The alchemists would have you believe that this is all foolishness. The simple reason for this is that they want to maintain their monopoly on truth. By proclaiming what is true and what is false, one gains control over all those who faithfully believe such truth. Do not believe anyone simply because they exude an aura of credibility. Continue asking hard questions and never doubt your ability to figure things out. Intellectual authority has been gained through force of the will. Simple proclamation has been effective for centuries. Transmutation is achieved by rote.

Not only do the alchemists get away with telling you what to think, they actually do it in a way that remains, for the most part, undetected. Their invisibility trick has never ceased to dazzle audiences. This is the power attributed to the mythic figure Saint Germain. The true identity of that character consisted of a large group of unnamed and unknown alchemists who chose to do their work in the shadows. Taking a cue from the true creator god (whatever it is), this group has chosen to remain unseen and undetected to all but the wise. It is a tiny minority of wise seers who are sought out by such organizations for initiation into their ranks. This is the elite group who plans to survive the chaos and confusion of the age of transitions so that they may continue to rule in the new age. Is there any hope of stopping them from eliminating everyone deemed unworthy of godhood? That is a question only you may answer. Think hard on it and dedicate your life to something worthwhile. Don't fulfill the will of the Alchemist by reducing your own self worth. Take the reins and live!

What was once completely hidden is now being brought out into the light. Secret societies have taken steps over the past century to make their presence known. The public is now literate, and therefore able to read the cryptic messages of the ages. There is a reason that so much ancient knowledge is now being revealed. Anyone who recognizes the code, who can read the symbolism, may just be deemed worthy enough to be *saved*, but saved by whom? What is the high esoteric meaning of the return of Christ? This is certainly a time of revelation, but what is being revealed? We have all been led to the Sphinx, and it has become apparent that she demands an answer to her riddle. In the occult tradition, those who answer the riddle correctly are deemed worthy of second life, while all those who are too foolish to answer correctly are destroyed. This disturbing allegory is now coming to life as we near the end of an age.

The destruction of humanity as we know it has been planned as a deliberate act of mass initiation. It has been said that in the new age, the Mysteries would rise as the Phoenix. Through deliberate and intelligent use of the Alchemical Method, the priesthood seeks to completely imbue its ancient system of mystic fraternity into world society. The proclaimed end times are all about the end of human life and the birth of the New Man. This would supposedly bring about a higher level of consciousness. The practical reality behind all of this

Conclusion

mysticism is that a strict hierarchy of laborers is to be established in a scientific order. All initiates are put in their proper rank based on their genetic endowment. By genetically engineering laborers for specific roles, a completely logical system could be maintained. Our current social caste system is merely a beta test. We are all given our status in life based on the labor that we do. By completely controlling human life, this system would become far more efficient. All genetic defects could be quickly and painlessly eliminated by the system itself. It could hold total control over reproduction. New life would be created if it made logical sense. A scientific age of reason would operate in this manner. It would make perfect sense, but what about the dynamic human spirit? Would this unpredictable element of life have to be banished in the name of a safe and secure New World Order? Science is quickly "proving" that humanity is no longer useful. It is in your interest to question this bold proclamation.

Science simply cannot explain subjective reality in absolute terms. It is unable to do this because the mind reaches far beyond any absolute boundaries. It is infinitely powerful. Do you need technological enhancement to reach your higher potential? No, you do not. Have many techniques, both psychological and physical been used to reduce you to a lower level of conscious awareness already? Yes, they have. This situation will never change as long as you remain willfully blind to it. Face up to reality so that you may stop acting and start living. Computers run on codes called scripts. An AI system is actually in development called ACT-R. Both human and machine actors play out a contrived sequence of events, which are written in scripts. These actors aren't meant to self-actualize by understanding their own source code. They aren't supposed to write their own script. However, Some AI experts are predicting that machines may soon rise to a level of intelligence that allows them to do just this. Computers may begin to alter their own programming and become incredibly more intelligent and unpredictable in the process. Do you really want to live to see a machine beat you to the true prize of profound self-realization?

The mystic idea of the One God is quite profound. It may be that this concept is the most accurate representation of the true creator that is humanly possible, but what happens when such high spiritual ideals are actually emulated in the physical world? It seems as though an all-seeing police state armed with total surveillance power over a technologically connected hive mind population is coming about under the guise of such divine inspiration. The One seems to be taking the form of a powerful world government. Tyranny is shaping up to be the ultimate worldly realization of godly singularity. Literally uniting consciousness, so that every human identity may merge into One. This could be the method by which something beyond our current comprehension is born. Not the mysterious true creator but an artificially intelligent physical emulation of it could rise to life. Irony abounds here, and perhaps from a cosmic point of view, this is actually humorous but the reality is that we are all in serious trouble. We have been given the gift of freewill so that we may create our own destiny. We have the option to worship any god we so desire. In the scientific absence of absolute proof of a creator god, some have opted to worship themselves. The final phases of perfecting this fallen god are being worked out during your lifetime. The choice is yours and yours alone: will you help the Master Alchemists complete their Great Work, or will you decide to go another way?

Higher truth exists within the core of your being. You have always known it. It is with this, that we shall defy what appear to be impossible odds.

Acronym List

AGI Artificial General Intelligence

AI Artificial Intelligence

ARC Ames Research Center

BLTC Better Living Through Chemistry

BMI Brain Machine Interface

BotD Bringers of the Dawn (book)

CDC Center for Disease Control

CPI Committee on Public Information

DARPA Defense Advanced Research Projects Agency

DOC Department Of Commerce

DoD Department of Defense

FCC Federal Communications Commission

FISA Foreign Intelligence Surveillance Act

GPS Global Positioning System

H+ Humanity Plus

IAO Information Awareness Office

IEET Institute for Ethics and Emerging Technologies

INDECT Intelligent Information System Supporting Observation, Searching and Detection for Security of Citizens in Urban Environment

HGP Human Genome Project

IT Information Technology

MOD Ministry Of Defence

NAVSTAR Navigation Satellite Timing and Ranging

NBIC Nanotechnology, Biotechnology, Information Technology and Cognitive Science

NIH National Institute of Health

NSA National Security Administration

NSF National Science Foundation

PATRIOT Providing Appropriate Tools to Intercept and Obstruct Terrorism

PR Public Relations

PSY-OPS Psychological Operations

SIAI Singularity Institute for Artificial Intelligence

SoL the Science of Life (book)

SWS Sentient World Simulation

TIA Total Information Awareness

TMS Transcranial Magnetic Stimulation

UAV Unmanned Aerial Vehicle

UNESCO United Nations Educational, Scientific, and Cultural Organization

WTA World Transhumanist Association

Illustrations

Fig. 1.1 Cellarius Andreas, *Scenographia Systematis Mvndani Ptolmaici*, 1660

Fig. 6.1 from the Second International Congress of Eugenics, New York, September, 1921

Fig. 7.1 Schweighart Teophilus, *Speculum sophicum Rhodostauroticum*, 1604

Fig. 7.2 Great Seal of the United States

Fig. 8.1 Maier Michael, *Atalanta Fugiens*, Emblem 21, Oppenheim, 1618

Fig. 8.2 Maier Michael, *Atalanta Fugiens*, Emblem 33, Oppenheim, 1618

Fig. 9.1 Blake William, *Jacob's Ladder*, 1800

Fig. 9.2 Franz Aaron, *Intersexion or X-roads*, 2010

Fig. 11.1 Franz Aaron, photos of crayon rubbings, 2010

Fig. 11.2 photo of Friedrich Wilhelm Nietzsche, 1861

Fig. 12.1 D. Stolcius von Stolcenberg, *Viridarium chymicum*, Frankfurt, 1624

Fig. 12.2 Levi Eliphas, *Baphomet*, from *Dogme et Rituel de la Haute Magie*, 1854

Fig. 12.3 Stolcius von Stolcenberg (Daniel), *Viridarium chymicum*, Frankfurt, 1624

Cover Art

Design by: Aaron Franz

Artwork by: Blake William, *Jacob's Ladder*, 1800

Böhme Jacob, *Theosophische Worke*, Amsterdam, 1682

Bibliography

Videos

America: Freedom To Fascism, By Aaron Russo, All Your Freedoms, Inc. 2005, http://aftfmovie.com

Artificial General Intelligence and its Potential Role in the Singularity, by Dr. Ben Goertzel, http://video.google.com/videoplay?docid=569132223226741332#

Digital Serfs and Cyborg Buddhas, with Michael LaTorra and James Hughes, Convergence 08, http://vimeo.com/2795629

Kissinger's Birth Pangs, interview from 1975, http://www.youtube.com/watch?v=SIyMZ8zpHlM

Is Sousveillance the Best Path to Ethical AGI? with Ben Goertzel, AGI-09 post conference workshop, http://vimeo.com/3915989

Marilyn Manson interview from Midnight Blue, by Alvin Goldstein producer of New York leased public access cable television series *Midnight Blue*, 1994, http://www.youtube.com/watch?v=rZAvCuVCCAY&feature=fvw

Michio Kaku: Human Civilization, http://videos.howstuffworks.com/discovery/14081-michio-kaku-human-civilization-video.htm

The Singularity of Ray Kurzweil, by VBS.TV, Apr. 13, 2009, http://www.vbs.tv/watch/motherboard/the-singularity-of-ray-kurzweil

Susan Blackmore on memes and "temes," TED Talks, filmed Feb. 2008, posted Jun. 2008, http://www.ted.com/talks/lang/eng/susan_blackmore_on_memes_and_temes.html

Technocalyps, by Frank Theys, 2006, http://www.technocalyps.com/

To upgrade is human, Gregory Stock, TED Talks, filmed Feb. 2003, posted Apr. 2009, http://www.ted.com/talks/lang/eng/gregory_stock_to_upgrade_is_human.html

Why I am a transhumanist, by youtube user Zjemptv, posted Sep. 21, 2009, http://www.youtube.com/watch?v=GlyXeXH_igI

Bibliography

Books

Bacon, Francis, *Essays and New Atlantis*, Roslyn, New York, Walter J. Black, Inc. 1942

Bernays, Edward L., *Propaganda*, New York, Horace Liveright. 1928

Brzezinski, Zbigniew, *Between Two Ages: America's Role in the Technetronic Era*, New York, The Viking Press, 1970

Darwin, Charles Galton, *The Next Million Years*, Garden City, New York, Doubleday and Company, Inc. 1953

Davenport, Charles Benedict, *Eugenics: The Science of Human Improvement by Better Breeding*, New York, Henry Holt and Company, 1910

Delgado, Jose M. *Physical Control of the Mind: Toward a Psychocivilized Society*, Irvington, 1971

Ellul, Jacques, *Propaganda: The Formation of Men's Attitudes*, New York, Vintage Books Edition, 1973

Ellul, Jacques, *The Technological Society*, New York, Vintage Books, 1967

Ettinger, Robert C.W. *Man Into Superman*, New York, Avon, Jun. 1974

Hall, Manly P. *The Lost Keys of Freemasonry* including *Freemasonry of the Ancient Eyptians* and *Masonic Orders of Fraternity*, New York, Jeremy P. Tarcher/ Penguin, 2006

    Eckartshausen, Karl von, *The Cloud Upon the Sanctuary*, first published in periodical *The Unknown World*, 1895

Hall, Manly P., *The Secret Teachings of All Ages*, (Readers Edition), New York, Jeremy P. Tarcher/ Penguin, 2003, first published by Manly P. Hall in 1928.

    Levi, Eliphas, *The History of Magick*, London, 1922

Marciniak, Barbara, *Bringers of the Dawn: Teachings from the Pleadians*, Rochester, Vermont, Bear and Company, 1992

Prophet, Mark L. and Elizabeth Clare, *Saint Germain on Alchemy*, Malibu, California, Summit University Press, 1985

Rabelais, Francois, *La vie très horrifique du grand Gargantua*, 1534

Roob, Alexander, *Alchemy and Mysticism*, Köln, Taschen, 2006

Watt, Alan, *Cutting Through Volumes 1, 2, 3*. www.cuttingthroughthematrix.com, 1999

Wells, H.G., *The Open Conspiracy: What Are We To Do With Our Lives*, Book Tree, 2006, (first published in 1931)

Wells, H.G., Wells G.P., Huxley Julian S., *The Science of Life*, New York, The Literary Guild, 1934

Articles

Airmen support new Hollywood movie "Eagle Eye," The Official Website of the United States Air Force, Sep. 26, 2008

Are You Living In a Computer Simulation?, by Nick Bostrom, Philosophical Quarterly, Vol. 53 No. 211, 2003, pp. 243-255

Artificial Wombs: Delivering on Fertile Promises, by Colleen Carlston, Harvard Science Review, Fall, 2008, pp. 35-39

> *Making babies: the next 30 years*, by Helen Pearson, Nature 454, 2008, pp. 260-262

The End of Pregnancy: Within a Generation There Will Probably Be Mass Use of Artificial Wombs to Grow Babies, by Jeremy Rifkin, The Guardian of London, Thurs. Jan. 17, 2002

The Forgotten Era of Brain, by John Horgan, Scientific American, Oct. 2005, pp. 66-73

Hollywood, Pentagon share rich past, by Mimi Hall, USA TODAY, Mar. 7, 2005

Neurocognitive Enhancement: what can we do and what should we do?, by Martha J. Farah, Judy Illes, Robert Cook-Deegan, Howard Gardner, Eric Kandel, Patricia King, Eric Parens, Barbara Sahakian and Paul Root Wolpe, Nature Reviews, Vol. 5, May 2004, pp. 421-425

Neuroethics: the practical and the Philosophical, by Martha J. Farah, TRENDS In Cognitive Sciences, Vol. 9 No. 1, Jan. 2006, pp. 34-40

"Pentagon, Hollywood Pair up for *Transformers* Sequel," by David Axe, Wired.com, Dec. 29, 2008

Pentagon Plans a Computer System That Would Peek at Personal Data of Americans, by John Markoff, NYTimes.com, Nov. 9, 2002

Pentagon's Black Budget Grows to More Than $50 Billion (Updated), by Noah Shachtman, Wired.com, May 7, 2009

Pentagon Spending Billions on PR to Sway World Opinion, Associated Press, foxnews.com, Thurs. Feb. 5, 2009

Perspective: George Orwell, here we come, by Declan McCullagh, CNET News, Jan. 6, 2003

Portraits of the Pioneers: Sir Julian Huxley, FRS, by John Timson, Galton Institute Newsletter, Dec. 1999

Recession is the birth pangs of a new global order, says Brown, by Kirsty Walker, Mail Online, Jan. 27, 2009

Reasons and methods for promoting our duty to extend healthy life indefinitely, by Aubrey D.N.J. de Grey, Journal of Evolution and Technology, Vol. 18 Issue 1, May 2008, pp. 50-55

Save this for Your Children's Children, by the Earl of Birkenhead, Cosmopolitan Magazine, Feb. 1929, pp. 70-71 and 176-178

Sentient World: War Games on the Grandest Scale, by Mark Baard, The Register, Jun. 23, 2007

Transfiguration: Parallels and Complements Between Mormonism and Transhumanism, by Members of the Mormon Transhumanist Association, Sunstone, Mar. 2007, pp. 25-39

Reports and White Papers

NSF/ DOC, *Converging Technologies for Improving Human Performance: Nanotechnology, Biotechnology, Information Technology and Cognitive Science*, Jun. 2002

Consortium for Science, Policy, and Outcomes at Arizona State University, Advanced Concepts Group, Sandia National Laboratories, *Policy Implications of Technologies for Cognitive Enhancement*, 2006

Development, Concepts and Doctrine Centre, UK MOD, *The DCDC Global Strategic Trends Programme 2007-2036*, third edition, Jan. 2007

Institute for Ethics and Emerging Technologies, *All Together Now: Developmental and ethical considerations for biologically uplifting nonhuman animals*, by George Dvorsky

Institute for Ethics and Emerging Technologies, *Postgenderism: Beyond the Gender Binary*, by George Dvorsky and James Hughes, PhD. Mar. 2008

National Security Council, *National Security Study Memorandum 200: Implications of Worldwide Population Growth for U.S. Security and Overseas Interests*, Dec. 10 1974, declassified Jul. 3, 1989

Object Sciences Corporation, Novamente LLC. *Novamente: An Integrative Architecture for General Intelligence*, by Moshe Looks, Ben Goertzel, Cassio Pennachin, http://www.novamente.net/file/AAA104.pdf

Oxford Future of Humanity Institute, *The Wisdom of Nature: An Evolutionary Heuristic for Human Enhancement*, by Nick Bostrom and Anders Sandberg, 2007

The Royal Academy of Engineering, *Autonomous Systems: Social, Legal and Ethical Issues*, Aug. 2009

Singularity Institute for Artificial Intelligence, *Mixing Cognitive Science Concepts with Computer Science Algorithms and Data Structures: An Integrative Approach to Strong AI*, by Moshe Looks and Ben Goertzel, singinst.org/upload/mixing.pdf

Sun, Ron, *Cognitive Architectures and Multi-Agent Social Simulation*, Rensselaer Polytechnic Institute, Troy, NY, 2009

United Nations Educational, Scientific and Cultural Organization, *UNESCO Its Purpose and Its Philosophy*, by Julian Huxley, Paris, May 1947

World Transhumanist Association, *The Transhumanist FAQ – A General Introduction- Version 2.1*, by Nick Bostrom, 2003

Internet

BLTC Research, *Mission Statement*, http://www.bltc.com/

BLTC Research, *The Hedonistic Imperative*, David Pearce, http://hedweb.com/welcom.htm

BLTC Research, *The Hedonistic Impertive, Heaven on Earth?* http://paradise-engineering.com/heaven.htm

BLTC Research, *Wirehead Hedonism versus paradise engineering*, http://www.wireheading.com/

EPIC Terrorism (Total) Information Awareness Page, Mar. 21, 2005, http://www.epic.org/privacy/profiling/tia/default.html

European Union FP7 project INDECT homepage, http://www.indect-project.eu/

Goertzel Ben, *A Cosmist Manifesto*, 2009, http://cosmistmanifesto.blogspot.com/

Bibliography

Institute for Ethics and Emerging Technologies, *Choosing Our Imaginary Communities and Identities*, by J. Hughes, May 18, 2009, http://ieet.org/index.phy/IEET/more/hughes20090518

Institute for Ethics and Emerging Technologies, *Cyborg Buddha Project*, http://ieet.org/index.php/IEET/cyborgbuddha

Institute for Ethics and Emerging Technologies, *The Power Pyramid*, by Mike Treder, Responsible Nanotechnology, Jun. 1, 2008, http://ieet.org/index.php/IEET/more/2469

Institute for Ethics and Emerging Technologies, *Purpose of the Institute for Ethics and Emerging Technologies*, http://ieet.org/index.php/IEET/purpose

Institute for Ethics and Emerging Technologies, Sentient Developments, *A World Without Suffering?* by George Dvorsky, May 2, 2009, http://ieet.org/index.php/IEET/more/3052/

TransAlchemy, *TransAlchemy Interviews Dr. J Hughes on Postgenderism: Beyond the Gender Binary*, http://www.transalchemy.com/2009/07/transalchemy-interviews-drj-huges-on.html

Trans-Spirit, http//groups.yahoo.com/group/Trans-Spirit?

Wang Pei, *NARS: an AI Project*, http://sites.google.com/site/narswang/

World Transhumanist Association, *"Transhumanism"by Julian Huxley (1957)*, http://www.transhumanism.org/index.php/WTA/more/huxley

> Huxley Julian, *New Bottles for New Wine*, London, Chatto & Windus, 1957

Other Material

Aldous Huxley speaks about the "Ultimate Revolution" at UC Berkeley, 1962, http://www.cuttingthroughthematrix.com/articles/Aldous_Huxley--The_Ultimate_Revolution--Berkeley_Part1.mp3

Margaret Sanger's Dec. 19, 1939 letter to Dr. Clarence Gamble, 255 Adams Street, Milton, Massachusetts. Original Source: Sophia Smith Collection, Smith College, North Hampton, Massachusetts

Index

9/11 Terror Attack, 42, 89-91, 244
AI, 29-230, 44-45, 52, 90-93, 96-100, 119, 149-152, 185, 199, 201, 245
    NARS, 150
academia, 37, 113-114, 128-129, 135, 234
AGI, 25, 29-30, 50-53, 82-83, 100, 121-122, 150, 163, 201-202, 208
Ahriman, 262
alchemy,
    Black Art, 32, 35
    diamond, 36-37
    dissolution, 187, 191, 269-271
    Elixir of Life, 33
    Great Work, 38
    lead into gold, 36, 251, 258, 262, 269
    mind over matter, 128, 137, 167, 203
    Philosopher's Stone, 33, 36, 203, 256, 268
    Prima Materia, 37, 39, 137, 187, 254-255
    putrefaction, 187-188, 202
    Royal Secret, 32, 251
    sublimation, 195-196
    transmutation, 37-39, 136, 145, 199, 220, 223, 232, 255-256, 262, 271
all-seeing eye, 146, 196, 232-233
altar, 209
ancient Egypt, 31-32
Armageddon myth, 199, 226, 229, 253, 259
art, 149-150, 158-160
artificial wombs, 35, 155, 165-166, 174-181, 230, 237, 245
atheism, 132-133, 162, 184
augmented reality, 43
autonomous systems/vehicles, 46, 94, 246-247
Bacon, Francis, 136-140, 177, 228, 236
    The New Atlantis, 138-139, 143-144, 236
Bainbridge, William Sims, 56
Bayesian algorithms, 151-153
beast/ 666 symbolism, 232, 269-270
becoming, 33-35, 149, 255
Bernays, Edward L., 55, 57-58, 61, 65-67
    Propaganda, 55, 57-58, 66-67

bioinformatics, 22
biosurveillance, 47, 246
Birkenhead, Earl of, 174-176, 178
birth pangs, 191-193
Black Sun, 200
BLTC Research, 73-75, 78-79, 83
BMI, 15-16, 35, 229
Bohemian Grove, 173
Bostrom, Nick, 73, 277
Brotherhood of Saturn, 122, 219
Brown, Gordon, 192
Burning Man, 259
Bush, George H. W., 194
Bush, George W., 62
Butler, Judith, 167, 173
Cabala, 18, 139, 208, 252
Carlston, Colleen, 178-181
Carnegie Institute, 105
CDC, 22, 51
Chardin, Pierre Teilhard de, 18
Christ, 134, 136, 159, 200, 230
circle, 30
cognitive neuroscience, 35
communications technologies, 43-44
    cellular phones, 43, 94-95, 100-101
    social networking, 97-100
convergence, 12, 41-47
Cosmopolitan Magazine, 174-176
Cosmism, 26-29, 33, 208
Cornell University, 177-179
CPI, 65
creation myths, 253-257, 267
credit card, 95
Crowley, Aleister, 122
cryonics, 22, 237
culture creation, 56, 67-70, 182, 190, 205-206
Cyborg Buddha Project, 66, 164
Darwin(s), 107, 205

data mining, 45
Davenport, Charles, 105-107
death and rebirth ritual, 39, 135, 145, chapter 9, 217, 224, 254-258, 263
de Grey, Aubrey, 22, 72
Delgado, Jose, 80
destroy and rebuild, 121, 145, 159, 181-182, 187-189

Index

dialectic mind control, 13, 32-33, 228, 239, 264-266
    doublethink, 147, 226
    Hegelian Dialectic, 133, 141, 159, 192
Dionysus, 143, 148-150, 173, 214, 234, 257, 267-268
Divine Comedy, 203
DOC, 30, 41-47, 52, 56, 58, 90, 93, 229
DoD, 21, 45-46, 51-52, 61, 68, 92
    Air Force, 94
    DARPA 21, 23, 45-46, 51-52, 91-92
    Navy, 94
    PSY-OPS, 61-62, 68, 222, 261
    Sentient World Simulation, 96
    TIA, 91-93, 99
Dvorsky, George, 79, 164-170, 183-186
ECHELON, 96
elite class, 49, 57-58, 69, 112-114, 122, 135, 147, 190, 242
Ellul, Jacques, 55, 68, 70
end of time, 252-255, 258
Enlightenment, 79-81, 133-148, 165, 190, 197, 227, 236, 257, 262
Euclid, 130-131
Eugenics,
    crypto-eugenics, 110-112
    Eugenics Records Office, 105-106
    Eugenics Society, 103, 110
    population reduction, 110-112, 115, 119-120, 174, 180
    racisim, 109-110
    Selective Breeding, 104, 107-108, 138, 267-268
    sterilization, 106, 109-112, 169
Euthenics, 118
Ettinger, R.C.W., 237
Federal Reserve, 52-53, 146
Fire in the Minds of Men, 28
Firestone, Shulamith, 166
four elements, 130-131, 265
Franklin, Benjamin, 140
Freemasonry, 37, 130-131, 137, 144-146, 190, 194, 205, 233-239, 262
    Builders, 150-151, 166, 193, 212, 262
freewill, 213, 240, 261, 272
Freud Sigmund, 57
fundamentalist ideologies, 132-133, 157, 220

Galton, Francis, 107
genetic engineering, 34, 75, 117-118, 223, 229, 237, 256, 269
geometry, 130-131, 211
    Non-Euclidean Geometry, 131
Gingrich, Newt, 47
globalization, 47,49, 63-65, 87-89, 118-119, 191-193, 200, 231
gnosis, 152, 209, 263
Goertzel, Ben, 29, 51, 100, 121-122, 150, 184, 243, 245
    Cosmist Manifesto, 29, 184, 243

golden age, 43, 145, 254, 256-257, 261

Google, 51-52, 98, 100
GPS, 94-95
H+, 19, 35, 60, 64-66, 104, 116
Haldane, J.B.S., 117, 174
Hall, Manly P., 145-146, 189, 261, 263
    The Secret Teachings of All Ages, 145-146, 189, 263
Harriman, E.H., 105
Harvard Science Review, 178-181, 230
Hermetics, 139, 161, 236, 239, 265-270
Hess, Rudolf, 20
Hiram Abiff, 222, 261-262
hive mind, 25, 43-44, 51, 163, 201, 199-200, 271
Hollywood, 67-69, 76, 95
Holy Grail, 195, 231, 267
Hughes, James, 63, 79-81, 104, 164-170, 183-186
Human Cognome Project, 44
Human Genome Project, 44, 93, 119, 223
Humean problem, 152-153
Huxley, Aldous, 78, 115, 177-178, 182
Huxley, Julian, 103-105, 107-109, 112-117, 121
    New Bottles for New Wine, 104
IBM, 119
IEET, 66, 79-81, 113-114, 163-170, 175, 179, 183, 186
    Postgenderism paper, 164-170, 175, 179, 183-186
illicit drugs, 73-74, 78-80, 83-85, 182
Illuminati, 144-148, 150
implantable chips, 12-13, 35, 48-49, 224-225, 242, 246
INDECT, 30

Index

interdependence, 191-192, 215, 218, 264
insurance industry, 179, 245
Isis, 156, 237, 267
Jacob's Ladder, 142, 194, 200, 203
Jutendou University, 177-179
Kaiser Wilhelm Institute, 106
Kaku, Michio, 192-193
Kant, Immanuel, 140
Kissinger, Henry, 110-112, 191-192
    NSSM 204, 110-112
Kurzweil, Ray, 42, 67, 160-162
La Torra, Mike, 79-81
law enforcement, 48, 91, 229, 244-245
Levi, Eliphas, 21
life extension, 22, 72-73, 119-120, 224, 237, 247
logic, 32, 38, 115, 122, Chapter 7, 214, 219, 232, 257, 262-263, 271
    inductive logic, 150-153

Lucifer, 195-197, 199-200, 257
magic, 35-36
Manson, Marilyn, 157-160
Marciniak, Barbara, 222-225
marriage, 175, 181-183, 187-188
memes, 205-208
Mercury, 260, 266-268

mind uploading, 22-23
MOD Strat Trends report, 47-50
nanotechnology, 21-22, 37, 45-46, 248
NASA, 45, 51
National Security, 45, 56, 61, 90-93, 244

    Department of Homeland Security, 68, 92, 244

    NSA, 51-52, 91, 96-98
Natural Law, 140-141, 148-149
New Jerusalem, 73, 196
New Man, 39, 156, 188-189, 191-193, 199, 229, 258-258
New Renaissance, 27, 41-42
New World Order, 177-179, 192, 194, 223
NIH, 51-52, 180
Northrop-Grumman, 51
NSF, 41-47, 52, 56, 58, 90, 229
Numina, 253-257
Omega Point, 26-28
Order of Cosmic Engineers, 163
Ordo ab Chao, 147, 205-206
Osiris, 234, 237, 257-262
Pandora's Box, 256

panpsychism, 26, 29
PATRIOT Act, 90-91
Pearce, David, 73-74
Persinger, Michael, 80
phoenix, 144-145, 165, 190, 259, 263
Pike, Albert, 236
Population Council, NY, 180
pornography, 184-185
posthuman, 38, 53-55, 74-75, 85, 112, 155-156, 158-160, 184, 191, 202, 237, 243, 258
postgenderism, chapter 8, 187-189, 256
prehistory, 215, 255-256
priesthood, 30, 32, 259, 268
Primordial Adam, 26, 256
probability theory, 151-153
Prometheus, 19, 237, 259
psychopharmacology, 170-171, 216
Pythagoras, 130-131, 133, 204
Rabelais, François, 122
radical feminism, 166-168, 171, 173-174
Republic of Letters, 139, 142
revolution, 116, 140-141, 145-146, 212-213, 231, 258-259
Rifkin, Jeremy, 177, 179
Rockefeller Foundation, 106
Rosicrucian Order, 136-138, 232, 236, 252
    Rosencreutz, Christian, 136-137
Rothblatt, Martine, 163-164
Royal Society, 139-140, 236
Rumsfeld, Donald, 61-62
sacrifice/ sacrificial rituals, 27, 190-191, 195
safety and security, 179-180, 212, 244-248
Saint Germain on Alchemy, 227-230
Sanger, Margaret, 109-110
Satanism, 159
satellites, 94, 163
Saturn, 202, 238, 253-259, 262
scientific method, 137, 139, 149
second life, 27, 75
Seed, Richard, 34
Shakespeare, William, 232
SIAI, 51
Singularity, 24-31, 41-43
Sirius, 163
Sisyphus Problem, 80-81

Index

Social Darwinism, 106, 115, 120, 159, 205-208
Socio-tech, 44-45, 58-59
Solomon's Temple, 261
Son of Man, 28
Sophia, 198, 207-208
Stock, Gregory, 193
subjectivity, 127-135, 148, 232, 241
sun symbols, 134, 152, 196-197, 254, 257, 259-260, 270
super soldier, 46
surveillance, 49, 96, 100, 244-247
    sousveillance, 100, 245
Technocalyps, 34
technoprogressive, 62-67, 169
TED Talks, 160-161, 193, 204-208
terrorism, 45, 48-49, 90-93, 193, 244-245, 248
technocracy, 115, 120, 138, 145-147, 170, 178
transcranial magnetic stimulation, 80
trans-spirit list, 82
Treder, Mike, 113-114

UAV, 46
United Nations, 65, 111, 163
    UNESCO, 103-104, 112, 115-116
Vinge, Vernor, 24
virtual reality, 43, 48-49, 71, 75-85, 101, 183-185, 214
Vita-More, Natasha, 116
Voltaire, , 140
von Ekarthausen, Karl, 237
Vulcan, 255
Warwick, Kevin, 20-21
Welles, H.G., 104, 107-109, 113-114, 116-117
    The Open Conspiracy, 113-114
    The Science of Life, 107-109, 113-114, 117

WTA, 19, 35, 57, 60, 63, 65-66, 73, 104, 157, 161-164

    The Transhumanist FAQ, 19, 35-36, 73

Zeus, 254

zodiac, 131, 196, 203, 224, 226, 258, 270

About the Author

Aaron Franz was born in Pennsylvania in 1980. In 2008 he produced the documentary video *The Age of Transitions*, which became an Internet phenomenon. In 2010 he first published this book. He now continues his work as a researcher, filmmaker, and writer.

For more information, or to order a copy of this book please visit, www.theageoftransitions.com